DAVID DOUGLAS

A Naturalist at Work

DAVID DOUGLAS

A Naturalist at Work

An Illustrated Exploration Across
Two Centuries *in the* Pacific Northwest

JACK NISBET

SASQUATCH BOOKS
SEATTLE

This book is dedicated to people who work with plants of all persuasions.

Thanks to Steve and Karla Rumsey for their long-term dedication to a cause, and special thanks to Claire, Emily, and James.
This could not work without you.

∽

Printed in China

Published by Sasquatch Books
19 18 17 16 15 14 13 12   9 8 7 6 5 4 3 2 1

Cover illustration: Jeanne Debons
Cover design: Anna Goldstein
Interior design: Sarah Plein and Anna Goldstein
Interior design composition: Sarah Plein
Maps by Emily Nisbet

Library of Congress Cataloging-in-Publication Data
is available.

ISBN-13: 978-1-57061-829-1

Sasquatch Books
1904 Third Avenue, Suite 710
Seattle, WA 98101
(206) 467-4300
www.sasquatchbooks.com
custserv@sasquatchbooks.com

Some portions of this book appeared, in different form,
in *Columbia* magazine, *We Proceeded On*, *American Surveyor*,
and in particular the *North Columbia Monthly*.

CONTENTS

Portrait of David Douglas
*"Have you seen Douglas? I was greatly
impressed by his intelligence and modesty."*
—Dr. Thomas Stewart Traill

LIST OF MAPS

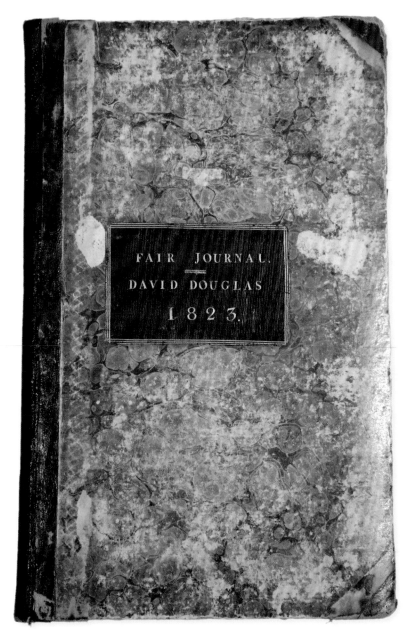

Journal from Douglas's 1823 collecting trip to the mid-Atlantic

PROLOGUE

Work in Progress

∽

"A tall splendid tree; leaves glaucous. The cones being on the top, I was unable to procure any. I went up one, but the top was too weak to bear me."

—*David Douglas, above Cascades of the Columbia, September 1825*

THE CONES OF EVERGREEN TREES—the seeds of most greenery, when you think of it—tend to grow at their most vibrant extremities. For members of the tribe of true firs, that usually means right at the crown. And since fir cones disperse their seeds by falling apart upon maturity, the only way to collect them at their moment of viability is to reach their exclusive domain. That's why I am ascending a grand fir, *Abies grandis,* on this breezy afternoon in early fall.

Grand fir
Abies grandis
"The bark of the young trees smooth and polished green, with minute round or oval scattered blisters, yielding a limpid aromatic fluid."
—David Douglas

Grand firs hold the darkest, shiniest green needles in the forests of my home in the Inland Northwest. Double-ranked on gracefully downswept branches, their whorls make ready stairs, and each of my scraping attempts to find a good foothold pours waves of sweet balsam through the air. After the first few steps, however, the limbs of my chosen tree bounce unnervingly, and I move close against the smooth-barked trunk for security. Blisters of clear pitch pop against my stomach as I corkscrew around the trunk, grasping for holds that become shorter and more constricted as the foliage closes in.

By the time I can swing my head back and spot the crown, the tree's leader is small enough to wrap a trembling hand around, and the top sways back and forth with each puff of breeze. On the uppermost star of foliage, a few small clusters of finger-length cones point to the sky. Each cone glistens with droplets of pitch that refract the late afternoon sun into shades of citrus green. Shoving through a last tangle of bristling new growth, I can almost touch them. Almost.

The leader seems entirely too thin to hold human weight, so I hug the tree for several minutes to consider the situation. My breath slows until it matches the easy metronome of the crown, until all my force flows downward through the trunk. That feels better. I wait, and slow my breath some more. On one forward swing I step up two whorls and stretch my arm to its limit, which is exactly enough to finger one gluey cone. I twist

it delicately, as if it were a hot light bulb. When the cone resists, I unscrew it a fraction more.

That brings a sound I have heard before, when stepping on a beetle barefooted at night. The cone disintegrates—no, explodes—in my hand. Brittle green sheaves, each one carrying a small fir seed, rain over my forearm. The pitch glues some of them to my sleeve. My forefinger and thumb close on a stringy central core—all that is left of my prize.

A second cone shatters the moment touch it. Unnerved, I retreat to my secure position and try to calm myself.

It is pleasant to be up here, rocking in the wind. As I stare at the glistening gems above me, I ponder how Scottish naturalist David Douglas reacted to this same situation on a similar fall day in 1825. After returning

Douglas squirrel
"I procured some curious kinds of rodents…which had been hitherto undescribed."
—David Douglas

to the ground, he tried to shoot some cones down with a musket, but the tree was too tall. He thought of chopping it down, but his hatchet was too small. He concluded a journal entry about his futile attempts with a note on the elusive seed: "Make a point of obtaining it by some means or other." And since Douglas was a persistent soul, within two years a London nursery was displaying grand fir seedlings sprouted from cones he had sent back to England.

For years, it seemed as though wherever I went, I could not escape from Douglas's persistent presence—from the withering chatter of his namesake tree squirrel to the smell of a particular wild onion; from the thrill of his blue clematis in early spring to the way his spiny short-horned lizard sat calmly in my palm. After following Douglas's tracks around the Northwest, I wrote a book called *The Collector* that attempted to tell the story of his adventures in the New World. When the book came out, so many people began to talk to me about their own experiences with the man and the country through which he had traveled that I realized I had only begun to touch the dynamic worlds he saw. So I went back on the trail, revisiting places he had described, checking on species of flora and fauna he had collected, following any lead that might reveal additional facets of his career and character. And wishing that I could get my hands on just one ripe grand fir cone.

———

David Douglas arrived at the mouth of the Columbia in the spring of 1825 with a very different purpose than that of the fur traders who greeted him. Board members of the London Horticultural Society and mentors in the British scientific community had charged him with expectations that ranged from the collection of showy garden flowers to promising new forest products and beyond, through all the realms of New World flora and fauna. Douglas carried letters of introduction from Hudson's Bay Company governors in London that would allow him to pursue these "treasures," as he called them, from the open Pacific to the crest of the Rocky Mountains.

Before leaving England, Douglas had pored over published works of the British and American explorers who came to the Northwest before him, and he soon learned to pay attention to the knowledge of company

Blazing star
Mentzelia laevicaulis
"Abundantly at the Great Falls of the Columbia, and gravelly banks of the River."
—David Douglas

laborers and local guides. He developed long-term relationships with several clerks and agents inside the Bay Company, as well as their mixed-blood families, who provided a conduit to the tribal knowledge of local flora and fauna. Douglas took advantage of these converging circumstances over the span of a decade in the Pacific Northwest, making three far-ranging visits before his sudden death in Hawaii in 1834. His days were spent in careful application of skills he had developed during a series of gardening apprenticeships, then honed while on a collecting trip through the mid-Atlantic states when he was in his early twenties.

Douglas's basic task was two-fold: first, to prepare pressed specimens, which would allow his cohorts back in England to fit the many unfamiliar plants he collected into their growing taxonomic catalog. His daybook entries show that he took great delight in beautiful blooms, and fretted about a shortage of the blotting paper essential for drying them. In the field his stacked blotters suffered damage during river crossings, grew mold and mildew, and were chewed on by insects and rodents. But all collectors

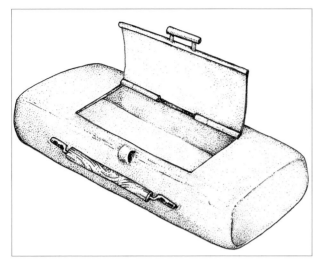

Vasculum

*Collecting botanists
stored plants and
seeds in tin vacula of
various sizes.
"I had in my pocket
my notes with my
small vasculum."*
—David Douglas

faced such obstacles, and Douglas's habit of gathering multiple examples yielded good results.

His second charge was to gather seeds and bulbs so that desirable plants could be propagated back in England. In order to time the collection of viable seed, he had to delve into the life history of each object of desire, and he took on the challenge with a flexible attack. He packed bulbs in sand and rolled up skeins of seaweed like scrolls. He calculated the pace at which cone sheaves slowly opened, and circled back to favorite plants over and over again in order to capture their seeds at exactly the right moment. He folded paper envelopes and stored favorites in his coat pockets for safekeeping. When it was time to ship them off for a nine-month sea voyage on which one splash of salt water could wipe out his entire effort, he cajoled ship captains to make sure they knew how to handle their fragile cargo.

Douglas's awareness extended beyond his appointed interest in plants to include every constituent of the landscape. He pinned insects and hammered rocks. He collected the eggs of the birds he was eating for supper, and the skins of the rodents that competed with him for ripe fruit and nuts. During a two-year sojourn in London, he tackled the steep learning curve of technical surveying in order to apply comparative geography to the range and relative abundance of individual species. All these activities required human assistance, and naturally drew him into the social dynamic of the Northwest at a time when cultures from across the continent and around the world descended on the Columbia, where families blended old and new lifeways throughout this diverse region.

The collector's meticulous daily routine, and his practical relationships with the people who helped him accomplish it, combined to broaden his impact. With every recipe for baked roots or identification of a plant fiber used for weaving, he illuminated tribal use of habitats over the millennia before European contact. When he recorded the exposure, soil type, wood

qualities, and durability of unfamiliar trees, his vision carried forward to the birth of a modern timber industry that has reshaped forests around the globe. When he stepped through worlds not quite his own—the fur trade, U.S. and European scientific communities, the British maritime empire, the politics of international boundary disputes—his observations provided a clarifying context.

Although the natural and human landscapes that Douglas described have endured a turbulent two centuries since his departure, a surprising number of the species he collected can still be found near the sites where he originally saw them. Pieces of our fragmented modern puzzle show up in his accounts as totem species and landmarks. At the same time, many details of both Douglas's and the Northwest's larger stories remain incomplete, waiting to be teased out of clues that have been left scattered behind.

Snow douglasia
Douglasia nivalis
"Upon close inspection, [Douglas] perceived that is was a genus entirely new. I have, therefore, named it after its indefatigable discoverer."
—John Lindley

Douglas Letter to DeWitt Clinton, 1825

"I soon found myself within the bleak dreary regions of Cape Horn."
—David Douglas

I.

WATERS OF THE WORLD

Crossing the Columbia Bar

✧

IN THE LATE WINTER OF 1825, the Hudson's Bay Company vessel *William and Ann* was eight months out of London. After rounding the Horn she had made two island stops, at Juan Fernandez and the Galapagos, before pointing north for the Columbia River. Her skipper, Henry Hanwell, had spent twenty years plying the company trade route between London and Hudson Bay, and this voyage marked his first taste of the Pacific. Off the coast of northern California, as the cook prepared the last of the tortoises they had picked up on the Galapagos Islands, the ship encountered the brisk equinoctial gales that blow down each spring from the Gulf of Alaska.

Albatross on Waves

"When rising from the water to soar in the atmosphere they partly walk on the water, tipping the surface with the points of the wings ere they can raise themselves sufficiently high to soar."

—David Douglas

Hanwell's single passenger, twenty-six-year-old David Douglas, was bound for the Columbia River to collect plants for the London Horticultural Society. He and the ship's surgeon, twenty-year-old John Scouler, shared a wide-ranging interest in natural history, and were thrilled to discover that albatross, which had been absent since the high latitudes of the southern hemisphere, reappeared with the blustery winds. They streamed baited hooks off the fiercely bucking stern, then dissected their catch to compare with the birds they had captured off the coast of Tierra del Fuego.

The farther the *William and Ann* wallowed toward her destination, the harder the relentless northwesterlies blew. Douglas, who once bragged of his seaworthy constitution, judged the "furious hurricanes of North-West America to be a thousand times worse than Cape Horn." Off the Oregon coast, blasts of spray breaking over the bow plastered small azure jellyfish (*Vellela vellela*) to the mainyard, which Douglas and Scouler peeled off for further study. In these northern waters the volume of their writings slowed

noticeably, however, and sometimes several days passed without a journal entry from either man.

On March 7th, the *William and Ann* reached the 46th parallel, only to be greeted by a large wave that stove in part of one bulwark. Captain Hanwell took in the foretopsail and pressed on, shipping a great deal of water. That afternoon, a break in the weather revealed the distinctive promontory of Cape Disappointment to the northeast, marking the mouth of the Columbia River. This welcome sight cheered John Scouler "with the anticipation that our voyage would speedily be at an end." But in the hours and days ahead, the cape lived up to its name. "Boisterous and frightful weather" continued, with relentless gales, frequent squalls of hail, and heavy seas. The carpenter fell on deck and hurt his leg, then suffered an epileptic attack; Scouler had to declare him unfit for duty.

The *William and Ann* remained at sea, tacking and wearing near the invisible line of 46° North, until the calendar page turned to April. At the beginning of the new month, two days of gentle breezes and a nearly calm surface lured Captain Hanwell close enough to the Columbia's maw

Ship in a Storm
"The weather was so terribly boisterous, with such a dreadfully heavy sea, that we were obliged to lay by, day after day."
—David Douglas

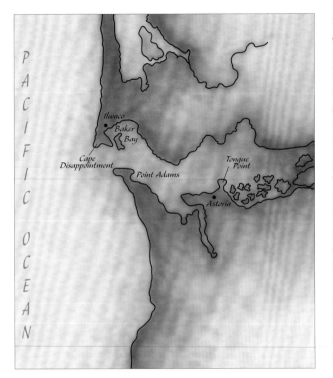

P
A
C
I
F
I
C

O
C
E
A
N

Ilwaco
Baker
Bay

Cape
Disappointment
Point Adams

Tongue
Point

Astoria

**Mouth of the
Columbia River**

to inspect the infamously cha-otic patch of water between Cape Disappointment on the north and Point Adams on the south. He carried the latest navigational charts and mariners' accounts of the river entrance, and would have been well aware of the hazards it presented. As the ship advanced within a half mile of the foot of the northern cape, with the tide beginning to ebb and the wind rising, Hanwell tacked and stood off. "The breadth of the river at its mouth is about five miles," Douglas later wrote. "The current is very rapid, and when the wind blows from the westward produces a great agitation. The water on the sand-bar breaks from one side to the other so that no channel can be perceived; when in such a state no vessel can attempt to go out or come in."

For three days the *William and Ann* huddled beneath close-reefed sails, waiting for conditions to settle. On April 7, surrounded by flying fog, they stood in once again. David Douglas recalled that "joy and expectation sat on every countenance, all hands endeavoring to make themselves useful in accomplishing this wished-for object." This time, even though the wind freshened and the haze turned to rain, Captain Hanwell pressed on, feeling his way toward a small opening in the breakers.

The Columbia River carries uncountable tons of sediment downstream to its wide mouth, where slackening current drops the finer grains in slithering skeins of sand and silt that form its estuary. The bar between fresh and salt water is further complicated by wide variations in river outflow, fluctuations of depth and temperature in the Japanese current offshore, and the interlaced schedules of moon and tide. The *William and Ann*, bearing a draft of eleven feet, had to negotiate this ever-shifting maze without the knowledge of a local pilot.

As the ship crept forward, Douglas and Scouler helped cast the lead to sound the depth of the channel. There were tense moments as those soundings guided the vessel around a bend that exposed her broadside to wind and surf, but in time, Douglas wrote, they "passed over the sand bank in safety (which is considered dangerous and on which, I learn, many vessels have been injured and some wrecked)." He later recalled the bar as the final test in a nine-month ordeal: "I arrived on the 7th of April. How would you like to be so long the sport of angry elements?"

On a spring evening very close to the week in April 1825 when David Douglas first entered the Columbia River, I step into the cabin of the pilot boat *Columbia*. The conditions are classic for this time of year: achingly clear sky, wind honking down from the northwest at twenty to twenty-five knots, a solid ten-foot swell from the west. We take off right into the breeze, heading down the river's wide estuary as a gray dusk chases the sun down over the vast Pacific. Between our boat and that glow, between fragrant marshland mud and a bottomless blue beyond the continental shelf, a distinct line of whitecaps licks across the interface of fresh and salt water.

Heaving the Lead
"Dr. Scouler and I kept the soundings, and safely passed over the sandbar, where many vessels have been injured and others lost."
—David Douglas

"I've seen those breakers rear up like a solid wall, from the south jetty clear across to the north, with no gap in sight," mentions the *Columbia*'s skipper, who is giving me a chance to ride in Douglas's wake for an evening. "It gives some people pause."

His vessel, a seventy-three-foot self-righting saucer, is especially designed to face these formidable conditions, and on this particular evening she is

North Head, Cape Disappointment

on her way to the other side of the white line to retrieve a bar pilot who is guiding a freighter out to sea.

We settle into the main channel and pass beneath the Cape Disappointment lighthouse, built in the 1850s. In the gloaming, the hump of the cape's basalt shoulder projects the same awesome power that Douglas felt on his initial sighting. Here, gazing up at the dark green forest atop the craggy outcrops, he caught his first glimpse of the tree that would one day bear his name.

Beyond the lighthouse point, we trace the shoreline's outward curve past a smattering of spooky tideline caves to the pocket inlet called Waikiki Beach, where an unfortunate group of Hawaiian sailors washed ashore after attempting to sound the north channel from a ship's dory back in 1812. This beach does not look like an inviting place for a swim.

Remnant of Shipwreck on Northwest Coast
"Did you hear of the total wreck of the Hudson's Bay Company ship on the sandbar at the entrance of River Columbia?"
—David Douglas

From the tumbled gravels of Waikiki Beach, we follow the massive boulder line of the North Jetty over two miles past the natural shoreline. As the population of the Northwest began to grow precipitously in the last half of the nineteenth century, the unruly waters at the river's entrance stood as a serious impediment to trade. In 1882 Congress authorized funds to make improvements on nature's design. A board of hydrographic engineers decided that narrowing the river's outlet would increase its velocity, so that the current would speed out to sea, carrying with it some of the sand and silt that had been clogging its mouth. Thirty years and several million tons of stone later, twin jetties reached out from Point Adams and Cape Disappointment. The carefully placed riprap did force the Columbia to scour a wider, deeper channel out into the Pacific, but could not stop the violent interplay between ocean and river. The wide bar, by every mariners' measure, remains among the most turbulent in the world.

At jetty's end we meet the chaotic patch of water that marks the bar. The skipper fingers his small twin throttles and plows through this tidal slop, his vessel twisting and pounding in a way that allows me to make full use of various grip handles arranged around the cabin. Although it's a relatively mild day, it takes little effort to visualize his description of other trips, when a little too much speed might cause his boat to leapfrog off the crest of one rogue swell, fly through the air, and submerge herself halfway up the next. Ratchet up any variable in the conditions, and we could quickly join the eighteenth-century mariner who wrote, "I never felt more alarmed and frightened in my life, never having been in a situation where I conceived there was so much danger."

Despite the dangers of the bar, moving between the Columbia and the open waters of the Pacific has always been an accepted risk of practical business. Only a few weeks after the *William and Ann* unloaded its cargo of fur trade goods and collecting supplies for David Douglas near modern-day Astoria, she exited the Columbia for a trading expedition north to Nootka Sound. When Captain Hanwell returned at the end of August, he stood off the bar for another day, waiting for another favorable breeze. Once safely inside, he was ready to receive the season's trade ferried down from Fort Vancouver.

Along with beaver pelts and a dozen other miscellaneous furs collected by the Hudson's Bay Company, the purser filled the ship's larder with a ton and a half of potatoes purchased from an Iroquois gardener. Carefully stowed in an airy situation above the *William and Ann*'s waterline were additional items of commercial and scientific importance, the gleanings of David Douglas's first season on the Columbia—bundles of dried plants, pressed specimens, bulbs buried in sand, all packed as carefully as possible in tin boxes and wooden crates. One chest contained a wealth of seeds, including new varieties of lupines, penstemons, lilies, monkeyflowers, evening primroses, and currants that Douglas hoped Horticultural Society nurserymen would be able to cultivate into commercial garden successes.

As word came of the *William and Ann*'s imminent departure for England, the collector worked frantically at Fort Vancouver to get his last bundles and letters dispatched in time. Hearing that she had been delayed by bad weather, he set out in a tribal canoe, roasting fish for the paddlers

Cape Disappointment
"The breeze improving as the day was advanced, we had the pleasure of entering in perfect safety this dangerous place."
—David Douglas

as he urged them to race downstream. They arrived only a few hours after Captain Hanwell, who had been fogbound for four days in Baker Bay, weighed anchor and "ran out against a very heavy Sea."

———

It's almost dark by the time we surge through the crossways pounding of the bar into something that resembles a directional swell, near the buoy generally accepted as the boundary between the Columbia and the Pacific. Of course, the same factors that make the bar so unpredictable assure that this buoy is no more than a general concept. Over the radio, we listen intently as the bar pilot delivers instructions for our rendezvous with the work-worn freighter he has just escorted across the bar.

The *Genius Spirit*, manned by a crew of Chinese mainlanders, has been upriver near Portland for several weeks, taking on a load of slippery brown potash mined in British Columbia. When the time was right for her departure, an experienced river pilot guided the ship down the Columbia to a designated spot above Astoria. There, he swapped places with a member of the Columbia River Bar Pilots. Established in 1846 to ensure the safety of ships, crews, and cargoes crossing the treacherous bar, the CRBP is the oldest continuously operated business in the Northwest. The pilot's job is to stand in the wheelhouse of any ship in the world, read the conditions, and relay instructions to a helmsman that will steer the vessel safely out to sea.

To carry out his part of the program, the *Columbia*'s skipper leaves the marker buoy and jets a mile west to a spot that, accounting for the ebb tide, should be favorable for the exchange. The *Genius Spirit*, which we have been tracking on the radar all the way along, soon looms above us. Although our skipper rates the four-hundred-foot freighter as small, close up she seems huge; I gauge that if she swallowed the 145 tons of the *William and Ann*, the red paint change that marks her load line would hardly move.

With *Columbia* angling in on her port side, the freighter slackens her steady course. Out of a dusk rendered almost black by powerful work lights on deck, crewmen appear along the rail to watch. As the vessels close, the ocean swells grow in size and intensity, and the exchange—a feat this bar pilot and this skipper have accomplished literally thousands

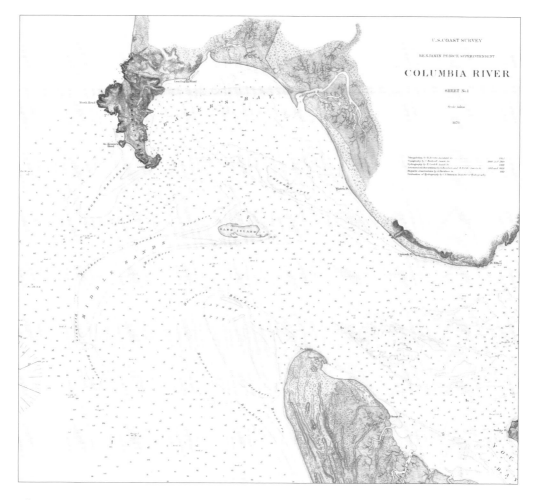

of times—seems anything but routine. The skipper comments, with eyes straight ahead, that *Genius* is too slight a vessel to create a solid lee and comfortable flat slick for him to settle into.

The absence of calm water does not alter the procedure. *Columbia*'s captain eases alongside the *Genius Spirit*'s hull, riding a swell that carries us up into the eyes of her staring crewmen, then drops us below the red paint line. Between these extremes, a hawser and rope ladder dangle from the freighter's deck rail. Gazing up from our position in the bottom of the trough, they look very far away.

Chart of Columbia River Bar, 1870
"The breadth of the Columbia is about five miles at its mouth; the current is very rapid, and produces great agitation."
—David Douglas

The skipper commits without a flinch, boring in on that textured hull as the next sideways chop tosses us into position. His deckhand clips on a harness and lifeline, shouting "ON DECK" as he steps outside the cabin. He works his way from stern to starboard to stand directly below the ladder, patiently waiting for the swell to carry him up so he can grip the manrope. The skipper touches his throttle and the *Columbia* growls into her appointed spot like a badger on the hunt.

High above us, the bar pilot appears as a ghost at the freighter's rail. In the midst of ocean chop and engine noise, his movements remain quiet and smooth. He descends the rope ladder, facing the hull, then chooses precisely the right moment to hop off the last rung and land solidly on the pilot boat's rapidly dropping deck.

The *Columbia*'s skipper eases off his dual throttles and pitches on the spot for a few minutes, surveying the situation. As the *Genius Spirit*'s klieg lights roll against the swell, the distance between the two vessels widens with surprising quickness. In the darkness of the cabin, bar pilot and skipper exchange a nod and one small comment on their evening's work: "A little tight back there, wasn't it?"

By the time the *Columbia* begins to make her way in, the freighter appears as only a dim glow in the distance. With a potash cargo silent in her hold, she has the breadth of the Pacific to cross before the deckhands assemble to drop another rope ladder off the side, signaling arrival at their next port of commerce.

During his decade-long career as a collector in the New World, David Douglas spent a total of more than two years as the sport of angry elements, and threaded the treacherous sandbars at the mouth of the Columbia five times. His second incoming passage must have been the one that carried the most anxiety, due to a disheartening news flash he received shortly before leaving London. "Did you hear of the total wreck of the Hudson's Bay Company's ship on the sandbar at the entrance of River Columbia, with the total loss of all on board consisting of forty-six persons on the 11th of March last?" he wrote in a letter. "It was the vessel Dr. Scouler and I went out in 1824, and the Captain was first-mate in that voyage." Indeed, the *William and Ann* had met her fate while crossing the bar.

Shortly after absorbing that disaster, the collector sailed from London aboard another Hudson's Bay Company vessel, the *Eagle*, under the charge of Lieutenant Robert Grave. With a hearty pun on his skipper's name, Douglas enjoyed an uneventful voyage around Cape Horn. But as the *Eagle* hove to over the bad banks of the Columbia bar, everyone aboard was greeted by the unsettling sight of her sister ship the *Isabella* fast aground, being pounded to pieces by relentless waves. "The ship which sailed with us was totally wrecked on entering the River, but I am glad to say no lives were lost," Douglas wrote. He realized that he might well have been assigned to that very ship. "Think what a plight would have been mine," he mused.

Less than six months later, in November 1830, Douglas made his first outbound trip across the bar, aboard the Hudson's Bay Company vessel *Dryad*, bound for Monterey Bay in California. The *Dryad* was full of fresh-sawn timber and salted salmon rather than beaver pelts, and Douglas was riding a new wave of interchange between the Columbia and the outside world. Although early mariners often emphasized that outgoing trips were just as dangerous as incoming ones, the *Dryad* slipped across without a hitch.

———

The sky above the space where the *Genius Spirit* disappeared spreads stars across infinite blackness, replacing the western horizon with points of light that float up and down with the swells. On the way back in, the *Columbia*'s skipper recalls the dozen years he spent as a commercial fisherman based in Astoria, motoring out into the Pacific after tuna and salmon, before serendipity offered him a chance to train as a pilot boat skipper. His job was won, he believes, when under challenging sea conditions he built a huge Dagwood sandwich in the galley of the old pilot boat *Peacock*, unaware that a couple of hardened veterans had their eyes on him. The next day, when one of the brotherhood went down from a sudden heart attack, they called him to the wheel.

Such camaraderie of men afloat brings to mind an incident described by Douglas as he lingered outside the river's mouth aboard the Bay Company's *Lama*, inbound from a stopover in Hawaii in 1832. As they waited for suitable conditions to run the bar, another vessel approached from the north. It was Douglas's old ship the *Eagle*, still under the command of

Robert Grave; better yet, Douglas's close friend Archibald McDonald was on board, reporting back to Fort Vancouver after a season on the lower Fraser River. Undaunted by the rough swells, Archie McDonald hopped into the dory that was exchanging mail between the two ships, and was soon clambering up on deck—exactly the kind of exchange, I imagine, that 1846 bar pilots would have made when leading early square-rigged vessels inside. Archie and his "old friend David" spent two nights tossing outside the Columbia's maw, exchanging news and jokes but saying not one word about nasty weather.

Taking a cue from the skipper, the bar pilot spins out part of his own personal history. After running away to sea at age seventeen, chance placed him on the deck of a freighter bound for the Columbia. On his first trip across the bar, he recalls, the setting embraced him and never let go, even after he had worked his way up to become a commercial captain who steered tankers all around the world. He entered his name on the long list of candidates hoping to join the Columbia River Bar Pilots, and made sure he qualified. When a call came in some years later, he took the leap without hesitation.

The ebb tide has given way to slack by the time we enter the chaotic patch over the bar proper, and the ocean's dark rock and roll feels more like a caress than a threat. As we pass beneath the Cape Disappointment lighthouse and Baker Bay opens off our port side, the bar pilot explains how he sometimes imagines time flowing backward across these waters: back to the days before the jetties, when steady winds closed off the corridor from point to point. When the naval emissaries of Spain and Great Britain sailed unknowingly past the churning slot, while Yankee ottermen snuck inside. When Chinook and Cathlamet boatmen paddled their large dugouts between fresh and salt water, as if they could belong to both. It's clear that he feels part of a much larger tradition, absorbing knowledge from others who have faced this agitated boundary. Constantly crossing the shadow line between river and sea, between safety and danger, seems to have sharpened his outlook into an appealing mixture of curiosity, awe, and the practical focus of appointed work. It's a perspective not unlike that of David Douglas, who once remarked: "I am no coward either in the water or on the water, and have gazed unmoved, and even with pleasure, on the wildest uproar and tumult of the stormy days."

Encampment at Bakers Bay after the wreck of the Peacock, mouth of Columbia River

Encampment at Baker Bay

"Several canoes of Indians visited the ship and behaved civilly, bringing dried salmon, fresh sturgeon, and dried berries of various kinds."
—David Douglas

II.

GOING THEIR OWN WAY

The People of the Northwest Coast

∞

"THE SIGHT OF LAND was to me truly a luxury," David Douglas wrote from the *William and Ann*'s moorage in Baker Bay upon his arrival on the Columbia in 1825. Gazing from the ship's rail at the forested slopes of Cape Disappointment, he anticipated "the pleasure of again resuming my wonted employment." Douglas's employment at that moment involved wading into these exotic habitats in search of new vegetation. Before many days passed, he came to understand that he could enhance his chances of success by tapping into the knowledge of local tribespeople.

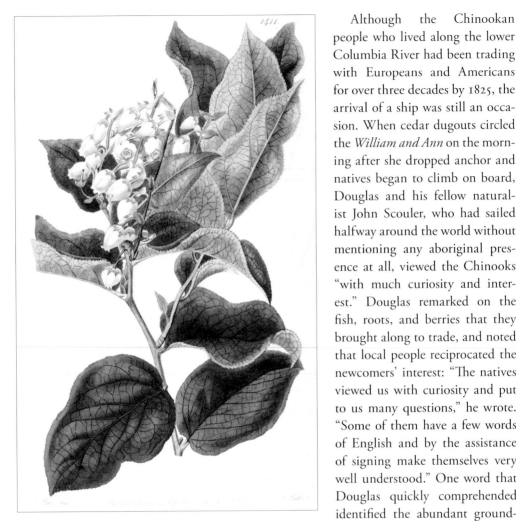

Salal
Gaultheria shallon

"On stepping on shore Gautheria Shallon was the first plant I took in my hands."

—David Douglas

Although the Chinookan people who lived along the lower Columbia River had been trading with Europeans and Americans for over three decades by 1825, the arrival of a ship was still an occasion. When cedar dugouts circled the *William and Ann* on the morning after she dropped anchor and natives began to climb on board, Douglas and his fellow naturalist John Scouler, who had sailed halfway around the world without mentioning any aboriginal presence at all, viewed the Chinooks "with much curiosity and interest." Douglas remarked on the fish, roots, and berries that they brought along to trade, and noted that local people reciprocated the newcomers' interest: "The natives viewed us with curiosity and put to us many questions," he wrote. "Some of them have a few words of English and by the assistance of signing make themselves very well understood." One word that Douglas quickly comprehended identified the abundant groundcover, *Gaultheria shallon*—a plant that Douglas and Scouler already knew of from conversations with Archibald Menzies, who had first collected it along the Northwest coast as surgeon and naturalist with George Vancouver's survey of the 1790s, and had particularly recommended it as a potential introduction to Great Britain. Douglas had also read its description by Meriwether Lewis, who rendered the Chinook word for the plant as *shallon*, which taxonomists had adapted as the Latin species name. Douglas, however, clearly heard local people call it *salal*. He recorded his own interpretation in his plant notes, and over time, this nugget of Chinookan nomenclature became the accepted common name of an iconic plant.

During that initial heady week of exploration, the Chinooks remained subjects of great curiosity and interest for the two naturalists. While Douglas focused on personal details such as clothing and decorative ornaments, the more educated Scouler drew some patronizing comparisons from popular books of the time: "These characters approximate them in some respects to the Mongolian race, with the Aetheopii they have manner of affinity." Before the first week was out, the physician had rudely barged into a lodge without an invitation, whereupon "the reception we experienced rendered it prudent to leave it as soon as possible."

The newcomers had a more congenial encounter with a group of Chinook women and children, whom they found gathering freshly sprouted shoots of the same field horsetail the naturalists knew from Scotland. As a taste treat, the gatherers nipped off the spearpoint-shaped tips, just as Europeans did with asparagus. Scouler found them quite palatable, and Douglas followed what today would be called this ethnobotanical bent when he recounted the way a group of tribal canoe paddlers snacked on young salmonberry shoots.

Chinook, Columbia River
"We met a number of Indians in the wood, chiefly women and children."
—John Scouler

Interior of a Chinook Lodge
"Cockqua, the principal chief of the Chenooks and Chochalii tribes . . . pressed me to sleep in his lodge."
—David Douglas

The visitors were soon introduced to Chief Comcomly, also known as Madsu, who lived with his band close to Fort George, the Hudson's Bay Company post at present-day Astoria. John Scouler wrote a perceptive description of the Chinookan headman and his village during two brief stopovers at the river's mouth, but Scouler returned to England at the end of the summer. Over the next two years, it was Douglas who became better acquainted with Comcomly and his network of Chinook family bands. On the naturalist's several excursions to the coast, Comcomly's people generously supplied him with food, shelter, transportation, guidance, personal introductions, and detailed information about local natural history. During these visits, Douglas found that "the natives are in general very friendly more so indeed than I was led to believe."

In July 1825, Douglas left his headquarters at Fort Vancouver with
a Canadian and two Indians to
continue his explorations on the
lower river. For two weeks the
naturalist and his guides worked
their way north across Cape
Disappointment, then around
Willapa Bay. In addition to ripe
seeds from the desirable salal,
Douglas tabulated about fifty new
plant collections. Although it is
impossible to trace the exact course
of his journey—Douglas was never
much for noting landmarks, and
his travel times varied according
to richness of habitat—at some
point during his return, he visited
the village of an earlier acquaintance whom he called Cockqua.
Although this Chinook headman
had learned several English words

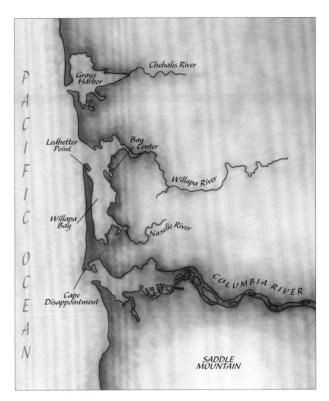

Pacific Northwest Coast

and phrases from seamen and fur traders, he probably communicated with
Douglas through Chinook jargon, a trade language that served as a lingua
franca throughout the lower Columbia. Cockqua treated his guest to a
fine sturgeon dinner, a welcome change from the collector's recent fare of
tea and biscuits. "I cannot but say he afforded me the most comfortable
meal I had had for a considerable time before, from the spine and head of
the fish." After supper, the headman invited the visitor inside his lodge,
where Douglas watched a girl weaving rushes, sedges, seaweeds, cedar
roots, and beargrass into some of the basic implements of Chinookan life:
"baskets, hats made after their own fashion, cups and pouches, of very
fine workmanship."

The next morning, the naturalist joined in a shooting contest with a
young man who threw his hat in the air as a challenge. Douglas wrote
that "my shot carried away all the crown, leaving nothing but the brim.
My fame was hereupon sounded through the whole country, and a high

value attached to my gun." Although somewhat vain about his skill as a marksman, Douglas was equally interested in his target. Upon examining the hat, he noted that it contained woven beargrass leaves, "which the Indians of the Columbia call QuipQuip, and on my observing the tissue with attention, Cockqua promised that his little girl, twelve years of age, should make me three or four after the European shape."

Cockqua, a man of gracious manners, presented his own hat to Douglas as a gift, along with an assortment of woven work. Douglas requested that the young weaver incorporate his initials ("DD") into the crown of one of the hats he had ordered. He went on to purchase "several articles of wearing apparel, gaming articles, and things used in domestic economy, for which I gave trinkets and tobacco." The gaming sticks, carved spirea shoots tipped with bright orange beaver incisors, indicate that Douglas's collections were growing beyond plants to include cultural artifacts, which were subjects of great interest in Britain at that time. Before departing, the naturalist accompanied Cockqua to a patch of evergreen huckleberry bushes and asked his host to collect seeds for him after the fruit ripened.

That August, "faithful to his proposition," Cockqua appeared at Fort Vancouver with a large packet of the requested huckleberry seeds. He also delivered three English-style hats woven from native materials. "I think them a good specimen of the ingenuity of the natives and particularly also being made by the little girl, twelve years old," Douglas wrote. "I paid one blanket (value 7s.) for them, the fourth included." The two men shared a smoke, then Douglas gave Cockqua "a dram and a few needles, beads, pins, and rings as a present for the little girl."

Later that fall, Douglas made another trip to the coast in search of, among other desirable treasures, beargrass seed. On his way downstream with four tribal paddlers, their small canoe suffered considerable damage from a collision with a submerged tree stump, forcing the boat close to shore to avoid the wind-whipped swells on the river. Stopping for the night at the Chinook village near Fort George, Douglas explained his plight to Comcomly and asked for the loan of a sturdier craft. "This old man sent his canoe and twelve Indians to ferry us across the river to Baker Bay, which they performed with great skill, though we had the misfortune to be taken in a violent storm while in the centre of the channel," Douglas wrote. "I attribute our preservation to the strength of the boat and dexterity of the Indians; by which, though the sea broke repeatedly over us in the middle

of the channel, we reached the shore in perfect safety, but with the loss of a few pounds of flour and a little tea, all our provisions with the exception of a few ounces of chocolate which I had in my pocket."

    Douglas and a fur trade companion were guided on this adventure by a brother of Comcomly's known as Tha-a-muxi, or "The Beard," who "imitated English manners with considerable nicety." The naturalist was hobbled by a lingering knee infection, and the four-mile portage across Cape Disappointment only exacerbated the problem. The party spent a miserable two days pinned down by violent weather, surviving on the little chocolate Douglas had preserved, augmented only by kinnikinnick berries and the roots of wapato and seashore lupine dug by Tha-a-muxi. When the storm finally abated, they completed the journey to Tha-a-muxi's home village on Grays Harbor, where Douglas received "every kindness and all the hospitality Indian courtesy could suggest, and made

Willapa Bay

a stay of several days at his house." During his stay, Douglas was able to gather seeds of beargrass and seashore lupine, though not as many as he had hoped.

From Grays Harbor, Tha-a-muxi guided his visitor up to the headwaters of the Chehalis River. After a friendship-sealing shave, he left Douglas and his Bay Company escort at the summit of the divide over to the Cowlitz drainage. The two followed an appointed trail to the main stem of that river, where another headman called Schachantaway escorted them to the confluence of the Cowlitz and the Columbia. From there Douglas, still troubled with knee pain, made his way back to Fort Vancouver.

The naturalist described his twenty-five-day absence as a time "during which I experienced more fatigue and misery, and gleaned less than in any trip I have had in the country." That being said, he did manage to collect seeds from the pliable sedge that the Chinooks used as a base for their woven fabrications, and to compare seashore lupine with the "wild liquorice" that the Chinooks had traded to Lewis and Clark. "There is in the root a large quantity of farinaceous substance, and it is a very nutritious wholesome food," he remarked.

A little more than a year later, in early December 1826, Douglas found the late-season inactivity at Fort Vancouver grating on his nerves. "My time laying heavy on my hands, I resolved on visiting the ocean in quest of Fuci [seaweeds], shells, or anything that might present itself to my view." A group of Chinooks had come in to trade, and Douglas and a companion from the fur post hired the men to ferry them downstream. They put in at the site of old Fort George, waited for the wind to die down, and crossed to the north side. There they were battered by another storm whose crashing waves destroyed their canoe and sent them fleeing into the woods.

Next morning Douglas and his party visited a son of Comcomly called Casicass and borrowed another dugout that they paddled and portaged for two rainy days to Willapa Bay. From there a short day's journey led to "the house of my old Indian friend Cockqua, who greeted me with the hospitality for which he is justly noted." Unfortunately for Douglas, that hospitality included some late-season dried salmon, which gave him a case of violent diarrhea. After four days of rest failed to stem the flow, the naturalist, fearing that he had contracted a strain of dysentery, retreated upriver. Despite the quick turnaround, his second trip

provided him with specimens of
wild cranberry and Labrador tea.

Before the winter's end, in early
March 1827, Douglas returned
to Willapa Bay with Hudson's
Bay Company agent Edward
Ermatinger. This time the naturalist's arrival coincided with the
death of one of Cockqua's cousins, and funeral preparations took
precedence over Douglas's plans.
Even so, the magnanimous headman promised that his hunters
would deliver some small mam-

Chinook Tomb
*"An Intermittent
Fever has depopulated
the country."*
—David Douglas

mal skins that Douglas desired to Fort Vancouver in time to make the
season's outgoing ship.

When Douglas returned to the Columbia River in the spring of 1830, he
no doubt intended to make another visit to Willapa Bay in hopes of sorting
out all the things he had missed in his previous attempts. But that summer,
as he collected in the Blue Mountains, a malaria epidemic known to the fur
men as intermittent fever swept through the lower Columbia country and
decimated the tribes. "Not twelve grown up persons live whom we saw in
1825," Douglas lamented to Scouler. "Villages . . . are totally gone . . . The
houses are empty, and flocks of famished dogs are howling about, while the
dead bodies lie strewed in every direction on the sands of the river." None of
the coastal Chinook people who had helped make Douglas's early journeys
on the lower river so productive—Comcomly, Tha-a-muxi, Schachantaway,
Casicass, Cockqua, or the twelve-year-old weaver of quality textiles—ever
appeared in his writings again.

The Shoalwater Indian Reservation is tucked into a protective shoulder
of land on the north side of the entrance to Willapa Bay, identified on
George Vancouver's 1792 map as Cape Shoalwater. A casino and gas sta-
tion now mark the turnoff from State Highway 105 that leads to the tribal
administration buildings; the road continues past them on a southeasterly

Chinook Cooking Basket

Woven of twined spruce root with beargrass overlay. "In the lodge were some baskets, hats made after their own fashion, cups and pouches, of very fine workmanship."
—David Douglas

course through a checkerboard of tribal housing and beach cottages to end at a crab processing plant on Toke Point. The highway wasn't paved until the 1950s, and local tribal members are well aware that the long isolation helped keep their culture alive. During the winter months, when raw winds snap against empty vacation houses, the world that David Douglas saw remains very much in evidence.

On a January morning, there is a bite in the air, and parting clouds reveal a line of snow along the Willapa Hills. The forecast for later in the day calls for several more inches extending all the way down to sea level, but Tony Johnson, Education Program Director, and Earl Davis, Heritage Director, assure me that snow never lasts long here. Johnson's roots in the last 150 years extend to the other side of the bay with the present day Chinook Nation. Davis's roots for a similar time period are tied to the Shoalwater Bay reservation. Both families trace much of their heritage prior to this to the north and south shores of the Columbia River. Johnson is married to Davis's cousin, and in the local tradition they consider themselves close family.

The floor and walls of Johnson's office are covered with traditional craftwork, including mask blanks carved from clear cedar; baskets of all sizes in different stages of manufacture; and student drawings on the walls sporting black, red, and white designs in the distinctive coastal style. Raw materials to make more, including the sedge locally known as sweetgrass, cedar roots, and beargrass, wait their turn. For Johnson and Davis, each of these traditional materials serves as a marker in space and time.

With both those elements in mind, I have come to explore two possibilities: that Chinook culture might be able to shed light on some of David Douglas's movements, and that his journal observations might provide small clues to the surrounding landscape of two centuries ago. I carry a modest list of personal names and Chinook or Chinook jargon plant terms that he mentioned on his journeys to Cape Disappointment and Willapa Bay.

We begin with the people. After a few false starts that might have something to do with the way white visitors hear and pronounce the local language, Johnson nods at the mention of one headman. That single name, for him, clarifies the routes that Douglas traveled. With one click of local knowledge, Douglas's vision of a ponderous slog through gloomy forests and rainy fens sharpens into a tour of a well-integrated universe.

"It begins on the River," Johnson explains. In the world of braided sloughs and intertidal waterways that flow through Willapa Bay, the River always means the Columbia, one day's journey to the south. Every family here has a place on the River, a traditional location determined by rank and marked by its distance upstream from the River's mouth.

Earl Davis unfolds a large area map marked with tribal place names. The traditional sites wheel out like a graceful constellation: fishing areas, shell-gathering beaches, berry patches, hunting havens, caches of cordage and weaving materials, and travel routes. The resources around each landmark are described by name in the tribal language.

On the big map, the sloughs that define the route from Cape Disappointment to Grays Harbor remain clearly visible. Some families continue to make the trip along them every year in their dugout canoes. They call it "keeping the road open" because that's the old travel path, and David Douglas represents just one of a succession of travelers. What's important to the Shoalwater, then and now, is the culture of bay and river; the language variations among the people who live there; the gathering of sweetgrass and huckleberries; the big dugout canoe plowing across the time-worn path toward the central village of Bay Center every year. Within that framework, many entries in Douglas's journals take on new meaning.

Douglas, for instance, noted that he saw Cockqua's people roasting the rhizomes of sword fern (*Polystichum munitum*) on embers for food. In another entry, he wrote that the roots of bracken fern (*Pteridium aquilinum*) were "dried and eaten by the natives, when roasted." Johnson listens to both

Seashore lupine
Lupinus littoralis
"For eating, the roots are roasted in the embers, when they become farinaceous.
The vernacular name of the plant is Somuuchtan."
—David Douglas

entries, then politely points out that Shoalwater people cook the roots of bracken fern, and only bracken fern, by the same method today.

On another occasion, Douglas described how Tha-a-muxi and his kin gathered the robust roots of seashore lupine and beach peavine—both purple-flowered members of the pea family—and cooked them for food. Johnson notes that people around Willapa Bay today bake only the lupines. Such a discrepancy might have arisen from a slight confusion during one of the naturalist's brief, rain-soaked visits to the bay; alternatively, eight generations might have winnowed family recipes from two plants to one.

Oral traditions and written observations mesh more seamlessly when it comes to salmonberry shoots, seen early in the spring. Douglas described their abundance and common use around Chinookan villages, and the emerging stems are among the most common cultural foods still gathered today. The naturalist considered salmonberry a potential ornamental shrub in Great Britain, but was not shy about nibbling the shoots as well as the berries: "fruit large, oblong, yellowish-white, and well flavoured."

Many Shoalwater families still gather salal berries, which is not surprising—Douglas was right when he noted that the small shrub "bears abundantly, fruit good, indeed by far the best in the country." During one of his visits, he purchased some cakes of pounded salal berries, made by women in the vicinity. After his return to England, Douglas forwarded some of the cakes to William Jackson Hooker in Glasgow, to show his mentor an example of good native food.

While struggling overland from Willapa Bay to Grays Harbor with Tha-a-muxi in the fall of 1825, Douglas survived on "*Sagittarea latifolia* called by the Chenooks Wapatoo" (known by other tribes as Indian potato or duck potato, and by some wildflower enthusiasts as broadleaf arrowhead). Among all the edible roots, wapato remains Johnson's favorite, and it is no surprise to him that both David Douglas and Meriwether Lewis enjoyed it during their time on the coast.

We step outside the office to look at some more of the Shoalwater landscape. North Cove is an inlet off the north end of Willapa Bay defined by the Tokeland Peninsula on the east and Graveyard Spit to the west. A long protective dike that separates the tribal buildings from the cove supports a healthy stand of evergreen huckleberry on its inland side. A few berries

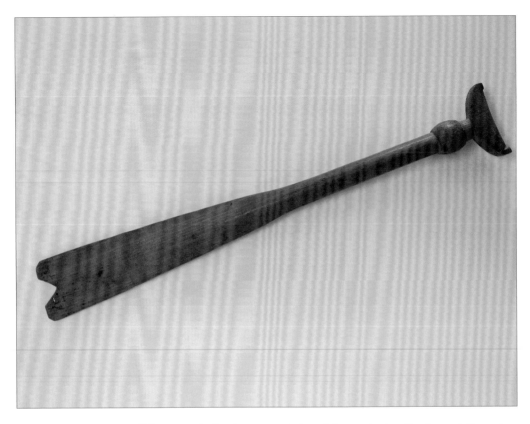

Chinook Paddle Made of Oregon Ash

"We had six Indians for paddling the canoe; they sat around all night roasting a sturgeon."
—David Douglas

still hang on the bushes, a reminder of the seeds that Cockqua delivered to Douglas at Fort Vancouver.

On the bay side of the dike, the morning's receding tide has exposed masses of sweetgrass, a resource still popular among basket makers across the region. These acres of sweetgrass snip the standard cord of time, hearkening back to the moment when David Douglas peered into Cockqua's lodge and watched a twelve-year-old girl plying her skill on cups, pouches, and baskets, one point of finger work along an endless skein.

The highway north past the tribal casino skirts the gray swells that lick across Willapa Bay's entrance and continues along the coast. Straight farm roads pointing inland connect a grid of commercial cranberry bogs, perfect squares that reflect deep red against the winter sky. The surface water leaking into this valley is controlled by ditches and pumps that distribute it among the scarlet bogs, but along the base of the hillsides, shades of the season's sleeping riparian growth appear. They mark an older slough that can still carry a paddler clear to Grays Harbor; among the native vegetation there, an interested gatherer can find the wild cranberry and Labrador tea that made two of Douglas's favorite collections. Local people call the latter simply "tea," and drink it with the same kind of enthusiasm that Douglas applied to his English black tea.

Back down on the Toke Peninsula, we pull off into a turnout and face south to study a line of perfectly displayed landmarks. A hill across the first broad inlet to the east hits the water at the important village site of Bay Center. Small bumps to the west mark four ponds south of Ledbetter Point, where other ancient villages disappeared after waves of fever and influenza. In the center of upper Willapa Bay rises the forested shape of Long Island, a place that in times past

provided families with refuge from such disease. South and west beyond Willapa Bay, Cape Disappointment stands as the last promontory at the Columbia's mouth. Tony Johnson counts them off, one by one, clear south to Saddle Mountain—the place of origin for local people, and a prominent peak in the Oregon coast range that still supplies good beargrass for basket designs.

Beargrass
Xerophyllum tenax
"The natives at the Rapids call it 'Quip Quip.'"
—David Douglas

Wrap Twined Basket

Beargrass wrap over sweetgrass sedge.

During his brief stops in the Chinook world, David Douglas visited many of these touchstones, and recorded hints of the cultural significance of the plants that captivated him. Along the way, he apparently came to realize how the coastal peoples were trying to hold on to that culture in the face of an expanding fur trade economy that was imposing itself over their world. "Most of the tribes on the coast (the Chenooks, Cladsaps, Cliketats, and Killimucks) are on the whole not unfriendly," he wrote. But "they are much prejudiced in favour of their own way of living."

III.

AWAKENING

The Roots of Plateau Culture

❧

JUST AS 1825 BECAME THE YEAR that David Douglas immersed himself in the landscapes of the lower Columbia and the coast, 1826 presented him with an opportunity to explore the Columbia Plateau—a vast region that included not only the arid shrub-steppe basin between Celilo and Kettle Falls, but also the surrounding high country featuring the Blue Mountains, the Okanogan Highlands, and the Palouse Hills. With a promise of lodging and support at the Hudson's Bay Company's three inland fur trade posts, he left Fort Vancouver in late March aboard the spring express, a pair of small, fast canoes manned by expert paddlers and designed to carry messages across the Columbia District.

View of Omack Lake, Colville Ind Reserve from trail on south shore

View of Omak Lake

"Here, the whole country being covered with snow, nothing could yet be done."
—David Douglas

When the brigade arrived at Fort Okanagan during the first week of April, the agent in charge of that post greeted guest passenger Douglas with warm cordiality. The paddlers had been working a fur trade schedule, beginning "always a little before day, camping at dusk." They had laid over briefly at Fort Walla Walla, providing the naturalist with enough of a view of the snow-covered Blue Mountains to plan a summer excursion into their heights. But as they hustled around the Big Bend, he had not even had time to examine the strange red-barked pines, some of immense size, that had sprung up as the Columbia kissed the foothills of the Cascade Range. He was further frustrated by the lateness of spring in the Interior. On some of the benches around Fort Okanagan, "the snow lies 3 to 5 feet deep. Here the whole country being covered with snow, nothing could yet be done."

Early spring snow still blankets the Plateau highlands, and my boots fill with ice as I slog along a terrace toward the Fort Okanagan Interpretive Center. The confluence of the Okanogan and Columbia Rivers forms a broad bay below the deserted buildings, which are closed for the season. Along the trail, I peer through a range finder aimed at the river's mouth and the site of the original Fort Okanagan, half a mile or more away, at the base of a finger of land that slumps down into the water.

As I work my way along the rocky slope that leads there, the snow cover lightens, dwindling to nothing around basalt outcrops with good southern exposures. The fresh bare ground is woven

Hudson's Bay Company Interior Trade Houses

through with muddy gopher diggings and matted by tussocks of bunchgrass and gray fall weeds. A few sagebrush buttercups peek between the rocks, but it is some time before I notice that a small biscuitroot called salt-and-pepper, *Lomatium gormanii*, is far more abundant. The white umbrella flower sprays of this parsley family plant are so tiny that I have to get down on my knees to see the purple anthers sprouting from each center, with a color so rich and liquid that it looks as if tiny sea urchins are creeping over the ground.

Nearby, clumps of arrowleaf balsamroot are far enough along to show finger-length stems that glisten with silvery down, each one capped with an equally hairy button-sized bud. Until those buds burst into flower, the slender stems provide a traditional early food source for northern Plateau peoples, who peel the down away in strips, then bite into a blue-green crunch of nutrition.

A few shoots of the much stouter biscuitroot that local people call chocolate tips, honoring its deep purple-to-brown flowers, peeks up between

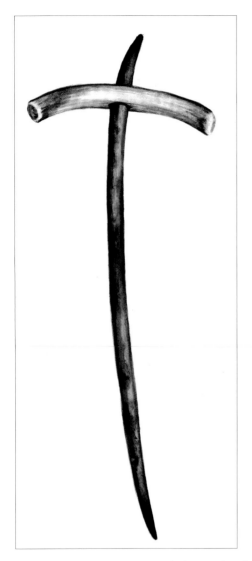

Digging Stick
Traditional Columbia Plateau digging sticks were often fashioned from hawthorne with an antler handle.

the balsamroot. Over the summer this potent *Lomatium dissectum* will grow large and somewhat toxic, but for the moment its first greenery offers another succulent treat.

Spring in the Inland Northwest often comes on fast, and by the time David Douglas reached the mouth of the Spokane River on April 10, the snow had receded enough for him to collect no less than five different species of biscuitroots, including two with purple anthers. He watched people make use of the same chocolate tips I saw at Fort Okanagan: "Umbelliferae, perennial; flowers purple; one of the strongest of the tribe found in the upper country; the tender shoots are eaten by the natives." Beside them he found the small lily now known as yellow bells (*Fritillaria pudica*). "This highly ornamental plant I must try to preserve roots to send home; roots eaten, both raw and roasted on the embers, by the natives and are collected in July and dried in the sun for winter use."

Because Douglas happened to be traveling with the Bay Company's spring express, he was perfectly positioned to observe all the plucking, digging, and picking, so similar to his own work, that carried tribal families around the region in carefully timed, endlessly varied cycles. Even though the collector recorded much information about several of the local food plants, he skipped over many other things in his journal. For example, he never described watching a group of Plateau families work their way across open ground, gathering tender green shoots. He never wrote about the flat twined bags that tribal women filled with biscuitroots and yellow bell bulbs, but he admired them enough to purchase one, woven from tawny Indian hemp, with a pattern formed from the deep purple basal shoots of beargrass. He never penned the phrase "digging stick" in his journal, but he must have marveled at the skill of women and children as they used their sticks to pry tender, deeply entrenched bulbs whole from rocky lithosols or the dense black peat of alluvial swales. In fact, he never described any tools he used to extract the perfect lily bulbs he packed in sand to ship back to England. The Columbia Basin's fractured basalt renders a gardener's trowel useless for this task, and a person in his shoes must have at least tried a digging stick, with its tempered hardwood or antler point angled for the purpose, a length adjusted to the height of the user, and a horn handle that allowed for subtle twists and powerful reefing.

It is only natural that Douglas's journals and plant lists reveal more about the collector than about the people who were actually living on the landscape. As Lakes (Sinixt) tribal member Marilyn James points out, every ethnographic detail he recorded would have reflected the traditions of the particular family he met on that particular day, and the single place where he happened to stop. Because of his ignorance of the language (Chinook jargon during this time was not in general use above Celilo Falls), he would have missed broad variations in the relationship between Plateau people and plants. Despite such limitations, his words do offer valuable snapshots of cultural life, including gathering times, preparation techniques, storage methods, and tastes, for he seems to have been game to try any new dish that was offered. While dining with a mixed-blood family at Spokane House, for instance, he sampled a new onion. Afterward, he requested that his hosts "should dry me the seeds . . . of a fine species of *Allium*, the roots of which were brought from forty miles above, on the banks of the Spokane River. Root as large as a nut, very pleasant and mild." Around Kettle Falls that same summer, he developed a fondness for another member of the onion family. "This plant is the only vegetable that I have to use in my food," he wrote. "I get it generally stewed down in a little dried buffalo-meat or game."

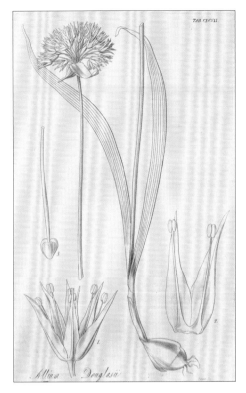

Douglas's onion
Allium douglasii
"A fine species of Allium . . . roots as large as a nut, very pleasant and mild."
—David Douglas

On a trip to the Blue Mountains, traveling with a Cayuse guide, the collector found yet another new species of *Allium*, with thin flat leaves and a tight mass of tiny, erect, superb pink flowers. William Jackson Hooker was sufficiently taken with the specimen that he named the plant "Douglas's onion." These blooms, so obviously a winner in nursery catalogs, still sweep in abundance along side slopes in the western part of the Blue Mountains, where they are still gathered by Cayuse, Umatilla, and Walla Walla families.

With his eye toward what might succeed in an English garden, Douglas devoted a great deal of time to the many other beautiful lilies that inhabit the Interior. The dramatic mariposas were of particular interest, with their obvious appeal for a horticulturist. He was aware that Meriwether Lewis had pressed a single specimen of elegant mariposa lily (*Calochortus elegans*) along the Columbia, but had gathered no seeds. Douglas not only obtained viable seed from Lewis's species, but also discovered two new members of the genus: the lavender sagebrush mariposa lily (*Calochortus macrocarpus*) and giant mariposa lily (*Calochortus nitidus*). Interested in their flavor as well as their beauty, he remarked: "in the spring the sagebrush mariposa forms an article of food in Inland Tribes, and is called in their tongue *Koo-e-oop* . . . The root is roundish, crisp, and juicy, yielding a palatable farina when boiled."

Douglas frequently compared tribal root offerings to farina, the finely ground cereal grains that generations of British children ate as breakfast gruel and dessert pudding. That porridge seems innocuous enough, but Douglas expressed stronger opinions about several of the roots he was offered. Western spring beauty (*Claytonia lanceolata*), well loved by many Plateau families, did not arouse him. "Its small roots are eaten by the natives, both in a raw state and cooked by roasting in the embers; when raw it is bitter and in every shape an insipid root." It appears that both Douglas and the tribal people he met shared a common trait: the dishes they enjoyed most were the ones they had eaten as youngsters. That is the definition of comfort food, whether it is a pudding spooned up in Perthshire or a roasted root popped into a mouth on a winter's night in the Colville country—both perfectly satisfying to anyone who has been eating them all their life.

Douglas did adjust his sensibilities for certain Plateau foods—taking advantage, for example, of the restorative powers of bitterroot (*Lewisia rediviva*). "The roots of this are gathered in great quantities by the Indians on the west side of the Rocky Mountains, and highly valued on account of their nutritive quality. They . . . are admirably calculated for carrying on long journies: two or three ounces a day being sufficient for a man, even while undergoing great fatigue." The collector's words echo almost exactly the sentiments of Spokane elder Pauline Flett, who marvels at the way a few strings of bitterroot provide one of her favorite tastes, as well as the energy to sing at stick game all night long.

The idea of a calculated scientist and a congenial dinner guest willing to adjust to the customs of the country come together in Douglas's two accounts of a lowland valerian. Known today as edible valerian (*Valeriana edulis*), it still may be found occasionally in swales along a traditional tribal trail connecting Kettle Falls with the Spokane River—a path Douglas traveled several times during the spring and summer months of 1826. This valerian first appeared as collection number seventy-three in his plant list for that year, an entry that pours out in a profusion of taxonomical, ecological, and tribal information.

> *Root large, fusiform, tastes somewhat like a parsnip; radical leaves entire (situated when in rich damp soils), lanceolate, smooth, three-nerved, floral pinnate, amplexicaul; flowers sulphur-yellow colour; plentiful in all low swampy grounds; the roots are gathered by the natives and boiled or roasted as an article of food (taste insipid); called by them* Missouii.

Three years later, William Jackson Hooker's *Flora Boreali-Americana* appeared with a more proper scientific description of *Valeriana edulis*, along with a personal aside from Douglas.

> *The roots, during the spring months, are collected by the Indians, baked on heated stones, and used as an article of winter or spring food. From a bitter and seemingly pernicious substance, it is thus converted into a soft and pulpy mass, which has a sweet taste, resembling that of treacle, and is apparently not unwholesome.*

Here Douglas sounds like someone who has sniffed the astonishing odor of a bright yellow valerian root as it was pried from the ground, who has watched the process of its preparation, and has gradually learned to

Colville Woman Digging Roots
*"Roots are gathered in great quantities by the Indians on the
west side of the Rocky Mountains."*
—David Douglas

like the soft pulpy mass fresh from an earth oven—what could be a better memory for a Scottish lad than sweet treacle, his ancestors' golden blend of molasses and corn syrup?

———

On a bright May afternoon, I crisscross the woods on a section of state land above the Colville Valley, near the trail where Douglas first collected edible valerian. Upon locating a patch of those plants, I extract a single large root: bright yellow, spindle-shaped, and very pernicious smelling. I walk back to my car and head for the town of Wellpinit on the Spokane Indian Reservation. Long before I get there, the odor has forced me to roll down my window.

Within an hour, I am standing with my root in the doorway of a room where several tribal elders are sitting around a table. I try to wait for them to finish their conversation, but before I can utter a word, loose hands begin to wave rapidly in front of noses.

"Whooo!" one of them says. "Where in the heck did you get that *msáwi*?"As the four women around her burst into laughter, I smile dumbly. The way she pronounces the Spokane word serves as a comment on the creditable job Douglas did with his phonetic spelling of "Missouii."

When I place the root on the table, their reactions begin to pour out. The word *msáwi?* refers to feet. No, not exactly, it means stinky feet. Like gym socks. Toe jam. And it tastes like that, too.

One elder remarks that her father would never let his own mother in the kitchen when she was carrying *msáwi?*—she had to keep it in her tepee behind the house. "But she'd rather stay out there and eat than come in the house anyway."

Another one's family never liked the stuff, and she never wanted to taste it—"How could you when it smells like that?"

A third elder recalls an uncle who used to store his baked *msáwi?* in an old blue Drum tobacco can. He kept it on a fence post, and she describes the round blue tin with the lid on, like it was going to explode. When he opened it, the smell of the black goo would make all the kids run away. "But you know," she says, "if you could just get past the smell, and that first taste, it was really good. We all liked it after we got used to it. It's like sweet jelly, but better, like real food."

Camas
Camassia quamash
"*Living on the roots
of camas.*"
—David Douglas

Even though he apparently learned to tolerate baked valerian root, Douglas would probably not have agreed that it was better than sweet jelly. A tribal food that received a more positive review from him was camas, although not all white visitors shared that opinion. Douglas, who often compared his own observations with those of explorers who came before him, noted another assessment of camas roots: "Captain Lewis observes that when eaten in a large quantity occasion bowel complaints. This I am not aware of, but assuredly they produce flatulence: when in the Indian hut I was almost blown out by strength of wind."

Despite certain side effects, camas constituted an important food on both the Plateau and the lower Columbia. Douglas's description of a baking method he observed offers a glimpse at one of the wide variety of camas recipes, many of which are still in use within Coastal and Plateau cultures today.

Its roots form a great part of the natives' food; they are prepared as follows: a hole is scraped in the ground, in which are placed a number of flat stones on which the fire is placed and kept burning until sufficiently warm, when it is taken away. The cakes, which

Camas Meadow

are formed by cutting or bruising the roots and then compressing into small bricks, are placed on the stones and covered with leaves, moss, or dry grass, with a layer of earth on the outside, and left until baked or roasted, which takes generally a night. They are moist when newly taken off the stones, and are hung up to dry. Then they are placed on shelves or boxes for winter use. When warm they taste much like a baked pear.

It is the first day of June. The air is wet with drizzle, and I'm standing beside Darlene Garcia, a Spokane elder, on a bunchgrass hummock in the channeled scabland country south of Spokane. We are watching a group of sixth-graders from the tribal school in Wellpinit wield digging sticks to gather bulbs of the blue-flowering lily *Camassia quamash*—the root that David Douglas thought tasted like a warm pear. Spokane people call it brown camas, because the roots turn a deep chocolate brown when they are baked. Local tribal people differentiate brown camas from white camas (*Lomatium canbyii*), whose white globes they dig in harsher scablands farther west in the Basin. Just as in Douglas's time, gardeners describe plants by the appearance of their flowers, while traditional Plateau people are more interested in their qualities as food.

The digging ground consists of several acres of rocky swales and rolling hummocks bounded by a highway bridge, a neck of ponderosa pine woods, and many acres of shin-high wheat. A railroad grade bisects the area into rough triangles, and the most noticeable plants on the scene are spectacular clumps of mule's ears (*Wyethia amplexicaulus*)—shiny, bug-covered leaf clumps that shelter bright sunflowers in full bloom. But the students are not here for wheat or mule's ears, and Garcia has eyes only for the day's harvest.

Members of her family dug roots in this area at least as far back as the 1930s, and old poles strung among the ponderosas hint at campsites far older than that. She grew up listening to stories about this place from her grandmother, who told her about patches of camas so thick that when they bloomed, people would mistake the expanses of blue flowers for lakes of standing water.

Back in the 1990s, Garcia decided to see if she could find the root ground that her family had worked. Another elder, who as a little girl had traveled to the place in a wagon, encouraged her. The spot was surprisingly easy to locate. Garcia matched features she had heard described—the neck of pine woodlands, the pattern of smooth regular hummocks so typical of these Palouse scablands, the way the railroad track cut through the shallow bowl—and stopped the moment she saw it. She climbed out of her car and looked around.

"I decided I would keep standing right there," she says, "and wait for somebody to come tell me that I was on their land."

The person who appeared was Chad Baxter, a retired pediatric surgeon who had purchased the land around 1970. From the former owner and various neighbors, he had learned that Indians used to dig roots around the railroad tracks, and he was curious to know more. He welcomed Garcia to his place and invited her to dig away.

"It wasn't easy at first," she recalls. "The bowl was thick with weeds, and the camas looked small and packed together. You could see a few big ones peeking up, and smell a few onions. I felt like the plants wanted us to start digging again. They just needed a little wake-up call. That's what Chad told me: 'You need to wake them up!'"

Garcia and her friends waited until the camas had bloomed and was going to seed, right around the beginning of June. Then they probed the slick black mud with digging sticks, fingering bulbs that ranged upwards from the size of thumb joints. They would snap off a swollen tuber, turn the plant's seedhead upside down into the hole, and heel it in. When they found many small plants bound together, they would turn up a clod and take the oldest, largest bulbs, then twist the clump back into the ground. Over a period of several years, the camas seemed to respond.

They found pink pungent onions (*Allium geyeri*) growing amongst the camas along one side of the depression, and a crescent ring of wild carrots or yampah (*Peridiridia gairdneri*) on a slight rise between the bowl

Douglas's brodiaea
Tritelia grandiflora
Brodiaea is a traditional root food on the Columbia Plateau.

and the patterned ground. They experimented with different families' methods for cooking all three species together in an earth oven. In time they started to bring students down from the school at Wellpinit to gather roots to be baked during the last week of school.

"We've been here with classes for four years in a row," says Garcia, as a student drops a handful of muddy camas bulbs into a five-gallon bucket. "We've done it enough that we can see how things change. When we pulled up in the car today, I stepped out and said, 'What is this? The camas are running away from us!' You could tell that the line of blue was a few feet farther back from the road, and we all thought they looked sparse. To me that means we have to go rediscover some of the other places. They say there used to be lots of other good digging sites right around this one."

Chad Baxter has passed on, but his daughter continues to watch over the digging grounds. She invites the farmer who has leased the land for more than forty years to drop by when the elders and students come to dig. They talk about the way the camas patch expands during particularly wet years, intruding on the wheat. They point across the big field toward a waving aspen grove, where a good-sized area of camas seems to be thriving in a wet depression. Beyond the railroad grade, there is a new clump that has been slowly expanding in size since it appeared about ten years ago. Would the digging be any good in one of those patches? It's a matter of following the seasons, of watching the snow melt away into standing pools around the rocks, of getting in tune with the sleeping and waking cycles of the camas bulbs as they wait underground.

Purple sage
Salvia dorii
"At Priest's Rapid on the Columbia, and plains from Walla-wallah to Spokan Rivers."
—David Douglas

David Douglas understood sleeping roots and new growth. After his snow-bound arrival in April 1826, he spent five months circling the Columbia Basin, checking in at Hudson's Bay Company posts for supplies, news, and guides. Although he often traveled with the traders on fur trade time, his search for fertile seed from plants that had caught his eye made his movements more closely resemble the annual round of a tribal band—he needed to arrive at a specific place to collect a certain resource at the moment it

was perfectly ready. There were many such moments for Douglas during his successive travels in the Interior in 1827, 1830, and 1833. He touched base at Fort Walla Walla no less than six different times, over a wide span of months. He was able to scour the ground for specimens on five different portages of Priest Rapids. He made a handful of separate forays into the Blue Mountains, and spent several productive weeks at Fort Colvile, walking out in different directions in the morning, collecting till dark, then stumbling back home to sort his treasures. More than a few times, he ate some of them for supper.

Flat Twined Bag
"An Indian bag of curious workmanship made of Indian hemp, [beargrass], and eagless' quills, used for carrying roots and other articles."
—David Douglas

Early 19th Century Field Telescope
Douglas carried a similar telescope in the field.

IV.

SCIENCE AND THE COMPANY

Outsiders in the Hudson's Bay Company Empire

❧

AS BRISK NORTHWESTERLY WINDS buffeted the *William and Ann*'s progress toward the Northwest coast in the early spring of 1825, Hudson's Bay Company governor George Simpson was visiting Fort George at the mouth of the Columbia. Simpson, recently placed in charge of the company's entire North American operations, was completing an overland tour of his domain. Since the 1821 amalgamation of the North West and Hudson's Bay Companies, the HBC had operated as a monopoly over the Columbia and Fraser River drainages, and Simpson was not pleased with the efficiency of the business. The governor therefore set in motion a series of changes focused squarely on the bottom line. He ordered the construction of a new regional command center, to be called Fort Vancouver, upstream on the Columbia opposite the mouth of the Willamette River. A large farm there would help to supply food for employees and reduce the costs of imports.

George Simpson
"Had a note from Governor Simpson."
—David Douglas

Although preoccupied with balance sheets, Simpson was an intelligent man well aware of the larger course of events. Before his departure from the Columbia, he penned a reprisal of his impressions, addressing both the state of the fur trade and relations with the local tribes, then turned his attention to a brief summary of the area's fauna and, more surprisingly, its flora.

The Columbia presents a wide field for botanical research as there is a great variety of Plants to be found every where . . . indeed, any one of experience in the study of natural history would add much to his store of knowledge therein by a visit to this part of the world.

By chance, just such a person was already en route for the Columbia River, aboard the Company's annual supply ship, the *William and Ann.* Had Simpson lingered at Fort George for another two weeks, he would have been on hand to greet David Douglas and read the accompanying letter from London headquarters explaining that the naturalist had come from England "for the express purpose of collecting Plants and other subjects of natural history" for the Horticultural Society of London.

The secretary of the Horticultural Society and the London directors of the Hudson's Bay Company were well acquainted, and the two organizations apparently saw mutual benefits in sponsoring Douglas's expedition to the Northwest. The HBC, having only recently assumed operations west of the Rockies, were anxious to learn everything possible about their new district, including information about its natural history. As the first trained naturalist to set foot in the interior reaches of the region since the Lewis and Clark expedition, Douglas faced a vast virgin collecting ground. The Hudson's Bay Company had since its inception supported scientific inquiry, and its governors were in the process of forming a small museum in London devoted to their territories in the New World, for which material from the Columbia would be most welcome. In fact, they had recently sent an appeal for specimens to their North American operatives, which Simpson had relayed to his traders spread across the region. "Do me the favour," he wrote to one agent in the polite language of a direct order, "to collect all the

seeds plants Birds and quadrupeds & mice & rats you can and let them be forwarded by the ship of next season."

As a guest of the company, Douglas would be under the care of John McLoughlin, director of the Northwest trade districts. McLoughlin had joined the North West Company in eastern Canada at age nineteen and risen through the ranks. Along the way, his somewhat domineering character gained enough respect that he was able to play a key role in the settlement of the bitter rivalry between the North West and Hudson's Bay Companies, and with so many Nor'westers crossing over to the HBC after amalgamation, he was the logical choice to head Simpson's new regime on the Columbia. A tall, dashing figure, McLoughlin appeared at Fort George a few days after the arrival of the *William and Ann*. He read the

John McLoughlin
"Mr. McLoughlin kindly sent me . . . one of his finest and most powerful horses for carrying my baggage or riding, which is of great service."
—David Douglas

naturalist's letter of introduction, then assured Douglas that he would render every assistance in his power. McLoughlin ushered his charge into a small canoe, whisked him upstream to the muddy construction site of the new Fort Vancouver, and installed him in a tent on the grounds of the post that would serve as the headquarters for Douglas's natural history endeavors off and on for the next nine years.

As far as the naturalist was concerned, McLoughlin more than lived up to his reputation as a cordial host. "I have all along experienced every attention in his power, horses, canoes and people when they could be spared to accompany me on my journeys," Douglas wrote. The chief factor responded in kind, assuring the London governors in his fall report that "in compliance with your directions we have given every assistance to Mr. Douglas which our means afforded and I am only Sorry, situated as we are, it has been out of our power to make him as comfortable as we would wish."

McLoughlin understood that making Douglas comfortable did not mean readjusting the schedule of the fur business, and his fall report made it clear that company interests came before Douglas's pleasure.

> He expressed a desire of going across the Continent in the Spring. But I informed him this would depend on the Instructions we would receive from York this fall as we might be so situated as not to have it in our power to accommodate him with a passage.

During his first summer at Fort Vancouver, Douglas took full advantage of McLoughlin's hospitality. While carpenters squared timbers for company buildings, the newcomer transplanted wildflowers and sowed turnips into a personal garden patch. He purchased necessary supplies from the company store, and hitched rides in canoes with employees traveling up and down the river. He filled first his tent, then a bark hut, with a fetid variety of plant, bird, and animal specimens. By midwinter, concerned about the soggy state of the collections in his hut, he accepted McLoughlin's invitation to move into the chief factor's half-finished living quarters.

Fort Vancouver
on the Columbia River

As Douglas became acquainted with the ways of the Hudson's Bay Company, he discovered that he shared a common background with many of the traders and clerks. The majority were, like Douglas, Scottish lads of modest means, second or third sons who stood no chance of inheriting their father's position or trade. For such young men, the fur business represented a means to strive for something more than they could expect in their homeland. Some of them found Douglas an amusing if somewhat eccentric companion. A number of these shared his interest in natural history and, after years of being immersed in the company's single-minded pursuit of pelts, were happy to engage in his far-flung interests.

As he moved through the country according to the patterns of the fur business, Douglas also came to know the lower ranks of workers. "I do not go alone," he wrote in an early letter home, "but in company with hunters when they go to trade with the Indians." These companions ranged

Fort Vancouver
"The scenery from this place is sublime—high well-wooded hills, mountains covered with perpetual snow, extensive natural meadows and plains of deep fertile alluvial deposit covered with a rich sward of grass and a profusion of flowering plants."
—David Douglas

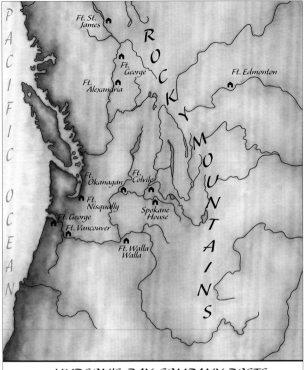

**HUDSON'S BAY COMPANY POSTS
PACIFIC NORTHWEST
1825 – 1833**

Hudson's Bay
Company Posts,
Pacific Northwest,
1825–1833

from French Canadian voyageurs to Iroquois, Eastern Woodland, and Plains tribal members who followed the trade, from numerous imported Hawaiians to the mixed-blood children of several generations of company employees. Douglas sometimes expressed frustration with the behavior of these laborers—there was the Iroquois man who drank the preservative rum from a bottle of precious salal berries, and a crew of voyageurs whose noisy paddle songs overran his ability to concentrate on his writing—but he tolerated the steady teasing that they directed at his curious habits. He also appreciated their wide range of outdoor skills, keen senses of humor, and distinctive personalities.

The workers, in turn, seemed to enjoy his energy and enthusiasm. A hearty traveler, Douglas could walk the expected twenty miles a day, and as an experienced marksman, he often helped company hunters lay in game. On several occasions the naturalist blistered his hands taking a turn with a paddle, and he considered himself a fair camp cook as well. He talked with factors, agents, clerks, free hunters, guides, translators, voyageurs, and anyone else with whom he could communicate, always searching out local information to complement his work. He took the jibes of company men as best he could, and tried to fit in—one of his most satisfying moments came in Oregon's Umpqua Valley when, after preparing a large kettle of venison and rice soup, he was able to feed a group of voyageurs who unexpectedly arrived in camp. "We have been entertaining one another, in turns, with accounts of our chase and other adventures," he wrote after the supper. "I find that I stand high among them as a marksman and passable as a hunter."

In return, several of these company acquaintances helped Douglas to further his collections of flora and fauna. John Work, the son of an Irish farmer, sent him seeds of antelope bitterbrush, and proved adept at preserving animal skins. When a mixed-blood hunter named Jean Baptiste McKay arrived at Fort Vancouver with a pouch full of roasted pine nuts that were far larger than those of any known species, Douglas supplied him with collecting bags and promised a reward for fresh seeds from that tantalizing tree. After camp dogs chewed up a pair of the naturalist's prized grouse skins at Fort Colvile, a distraught Douglas sent a message to agent Archie McDonald requesting replacements.

Temporary Camp
"I hastily bent my steps to my camp below . . . the storm raging still without the least appearance of abating."
—David Douglas

David Douglas was not the first scientific collector to depend on the Hudson's Bay Company: precedent for such support had been established during John Franklin's first northern expedition, from Hudson Bay overland to the Coppermine River, in 1819–22. At the beginning of that ill-fated journey, which saw eleven of Franklin's men die of starvation, an HBC governor sent an enthusiastic announcement to any company officers who might come into contact with Franklin's crew, encouraging them "individually and collectively to give this Gentleman every possible assistance he may require, in supplies, Provisions, men, dogs, Sledges, Hunters, in fact every thing that may be deemed necessary to render as facile as possible an undertaking so arduous and interesting." In a private letter, however, the same governor made it clear that while science might be interesting, the rights of the business had to be carefully protected, and company employees were ordered not "to make observations respecting the trade before the [members of the Franklin expedition] in conversation or otherwise."

Such official secrecy was naturally countered by loose talk around the trade houses, and it soon became apparent to many fur men that Franklin's expedition was something less than a juggernaut. After a conversation with Franklin's naval officer, Governor George Simpson decided that "there is little probability of the objects of the expedition being accomplished," mostly because of the lax physical habits of their leader: "Lieut. Franklin has not the physical power required for the labor of moderate Voyaging in this country; he must have three meals per diem, Tea is indispensable, and with the utmost exertion he cannot walk above Eight miles in one day."

This assessment proved correct, and from a company viewpoint none of Franklin's expeditions kept up the kind of steady pace through the country that had been required in the fur trade for decades. When Franklin's team planned another northern trek in 1834, company wags dismissed the effort as more of the same. "You'll learn what a fine story they'll make of this bungle," one agent wrote. "They will return next summer and like all the other Expeditions will do little and speak a great deal."

David Douglas knew nothing of John Franklin's limitations, but he had read the explorer's narrative describing his Coppermine expedition, and his lively account of sailing vessels, icebergs, aboriginal cultures, extensive natural history, and extreme hardship had obviously struck a chord. When Hudson's Bay Company traders from the Athabasca District carried the latest rumors of Franklin's second expedition back to Fort Vancouver in the late fall of 1825, Douglas excitedly wondered if he might somehow rendezvous with the explorer or one of his naturalists.

That wish was answered when Douglas accompanied a fur trade brigade east across the Canadian Prairies in 1827 and encountered one of the men he had hoped to see—assistant naturalist Thomas Drummond, who had

been collecting in the area for two years. Drummond happened to be at Fort Edmonton when a box of specimens Douglas had sent ahead arrived there. Seeing that the crate had obviously been dunked in water along the way, Drummond opened it to assess the damage. While acknowledging the kindness of this gesture, Douglas admitted that he could "not relish a botanist coming in contact with another's gleanings." Weeks later, when he met Franklin's chief naturalist, Dr. John Richardson, at another Saskatchewan River trade house, Douglas found himself awed by the quality and diversity of the physician's collections.

Moving down the Saskatchewan with the Columbia brigade, Douglas intersected the route of John Franklin at Norway House, on the north end of Lake Winnipeg. After a quick introduction, the explorer invited Douglas to join his canoe as he traveled south along the lake's eastern shore. For a lad from Scone village who had an admiring eye for celebrity, the ride with Franklin marked one of the high points of the trip. Douglas was no longer

Traverse of the Saskatchewan
"Sundry articles gleaned on my descent of the Saskatchewan River."
—David Douglas

working alone in the din of the fur trade, but traveling as a peer with one of the best-known explorers of his day.

Soon after that memorable canoe ride, Douglas made his way to Hudson Bay and sailed back to England. During the two years he spent in London, he worked with many eminent British naturalists and scientists. He delivered papers on the fruits of his research to meetings of the Horticultural Society, the Linnean Society, and the Zoological Society. He consulted with John Richardson about a book the doctor was completing on the fauna of northern North America; together, they visited the Hudson's Bay Company's fledgling museum to compare specimens of pocket gophers there with those Douglas had collected on the Columbia.

Nicholas Garry
"Mr. Work showed me a pair of Mouton Blanche of the voyageurs . . . but knowing them to have been procured at the particular request of Mr. Garry in London, I of course could not ask for them."
—David Douglas

During his visits to the company offices at Hudson House, Douglas developed a close friendship with deputy governor Nicholas Garry, one of the main supporters of the museum. In fact, while at Fort Colvile in 1827, Douglas had seen and coveted two beautiful mountain goat pelts that were being prepared for shipment to Garry in London. Described by other acquaintances as generous, open, and easy of access, Garry had a special interest in North America. He had traveled to eastern Canada in 1821 to implement the company's merger with the North West Company, and his diaries from that trip express his fascination with the people and the natural history of the region. He had accumulated a library of the accounts of travelers and explorers in the New World, and he would have been especially intrigued with Douglas's adventures in the Northwest. Garry likely also discussed with Douglas the impending boundary settlement between the United States and Great Britain, and its implications for the Hudson's Bay Company's posts on the Columbia. In a report that he wrote at this time for the Colonial Office, Douglas strongly supported the HBC's position that the boundary should follow the Columbia River from the mouth of the Snake to the Pacific.

A simple testimony to the nature of the friendship between the Bay Company administrator and the collector came after Douglas dropped by Hudson House in August of 1829 and learned some exciting news: "Mr. Garry was married on Tuesday last! He is off to the Continent. I do not know the Lady. He told me himself when I went to take leave of him on Saturday."

When the London Horticultural Society decided to send Douglas on a second expedition to the Northwest in 1829, the Hudson's Bay Company again "made a most liberal offer of assistance" and granted him free passage on their next supply ship to the Columbia. By the time of his return to the empire of beaver pelts, Douglas was realistic about the commercial priorities of the fur business, and had come to understand that few company employees shared Nicholas Garry's

Wavy-leafed silk-tassel
Garrya elliptica
"It forms a hardy evergreen shrub in the garden."
—William J. Hooker

expansive views. As he left Fort Vancouver for the last time in the fall of 1833, he considered the situation of Meredith Gairdner and William Tolmie, two young physicians who had recently arrived to serve in the Columbia District as both doctors and collectors. "These gentlemen have much to contend with. Science has few friends among a people whose only aim is gain on North West America, still I hope with such a person as John McLoughlin at the Columbia, a man can do a great deal of good indeed."

McLoughlin's boss George Simpson tersely summed up the company attitude when he learned later the same year that Dr. Gairdner was writing a narrative about his time on the Columbia. "I understand he is collecting materials for the press, and it is not desirable that our trade should be brought under public notice." In this respect the HBC was no different than many companies that shrink from outside eyes prying into details of their business. And although no similar complaints ever came to light about Douglas's own writings, Simpson would not have been pleased to know that Douglas had ignited a spat at Fort Kamloops when he told agent Samuel Black that "Company officers hadn't a soul above a beaver skin."

Many of the agents Douglas met during his travels had much more soul than a beaver skin, and developed lifelong friendships with the collector. One of those was Archibald McDonald, whom Douglas had asked to help replace his chewed-up grouse skins. Archie's mixed-blood son Ranald, who knew Douglas between the ages of one and ten, later recalled him as a close companion of his youth. John Work and William Kitson helped Douglas collect materials ranging from violet-green swallows to mountain goat fleece around Fort Colvile. Bay Company officials criticized agent Alexander McLeod for several of his decisions in the field, but he and Douglas often enjoyed an afternoon hunt on horseback together.

George Barnston, almost exactly the same age as Douglas, had grown up in the same part of Scotland. Originally trained as a surveyor and engineer, Barnston signed on as a clerk with the Hudson's Bay Company when he was twenty years old. He came west to work with the survey party that laid out Fort Langley at the mouth of the Fraser River, and shared quarters with Douglas at Fort Vancouver over the winter of 1826–27. Whereas George Simpson considered Barnston "touchy . . . and so much afflicted

Brown's Peony
Paeonia brownii
*"Flowers centre and the outside dark purple, on the edge and inside bright
yellow . . . flowering in perfection on the confines of perpetual snow; if in
my power, seeds of it must be had."*
—David Douglas

George Barnston
"I wish you had been with me to have lent me a hand."
—David Douglas

with melancholy or despondency, that it is feared that his nerves or mind is afflicted," Douglas quickly recognized a kindred spirit, and the two spent part of that winter trying to discern whether the California condors that soared over the lower Columbia used olfactory or visual methods for discovering carrion. Barnston, in time, would write as eloquently as any ornithologist about watching condors on the wing.

When Douglas returned to the Columbia in 1830 with newly acquired surveying skills plus a full set of instruments, the two forged an even stronger bond. With Barnston posted to Fort Walla Walla that season, they traveled together to the confluence of the Snake and Columbia Rivers, taking astronomical measurements of every description along the way. Douglas soon sailed to California, where he worked for the next two years, but the pair maintained communication through occasional letters. When Douglas came back to the Columbia in the fall of 1832, he found two carefully considered gifts from his Walla Walla friend waiting for him: a pressed Brown's peony, matching the collector's single most prized plant discovery, and a sample of kyanite, stunning blue crystals set in a mineral matrix that would delight any rock hound.

Upon receiving word of Douglas's death in a cattle pit trap in Hawaii two years later, Barnston wrote the most emotional response of any of the collector's acquaintances, from either the fur trade or the scientific world.

Unhappy Douglas! thy Shade is now before me, now shrieks for assistance—I yet hear the dismal and heart rending moans of my ill fated friend. Was there no hand to help? Could no Arm Save? Alas! the Almighty! his decrees are just and good, and the weary wanderer has been taken home. He had obtained a noble Conquest

*over warm and Powerful passions, his mind so often fixed on the
wonderful works of his Maker, had melted into love and good will
towards the whole creation, his spirit had caught the eternal flame
which dieth not, and he was hurried from a troubled seam to rest
in happiness & peace.*

Long after his friend's departure, Barnston held on to the interest in
natural history that Douglas had inspired. He continued to work for the
Hudson's Bay Company until 1863, moving to various posts with his wife
Ellen Matthews, the daughter of a Scottish fur agent and a Clatsop chief,
and at least some of their eleven children. All through that time, Barnston
collected insect cocoons and studied the larvae that crawled out of them.
When he and Ellen retired back to Montreal, he became an active member
of the Natural History Society of Canada, delivering papers on his insect
studies and serving as president of the Society in 1872–73. Eventually
one of his sons emerged as a botany professor at McGill University—one
member of a new generation of plant enthusiasts who could hearken back
to the traditions of the lower Columbia River, where one man's ardent
curiosity fueled the enthusiasm of a company world.

Nisqualey Half Caste Indians Gambling

Nisqually, Half Caste Indians
"I was hailed into the camp with 'Be seated at the fire, Sir,' and then laughed at
for losing myself in the morning, my game and other property in the evening."
—David Douglas

V.

INVISIBLE KIN

Mixed-Blood Families of the Fur Trade

❧

DAVID DOUGLAS'S JOURNALS and letters from the Northwest describe, for the most part, a world inhabited by men—tribal figures, fur trade agents, interpreters, hunters, and voyageurs. But his writings also hold hints of a more expansive reality: a squirrel-skin robe draped over the shoulders of a child; a girl with her hands full of weaving materials; boys selling fish or snaring lizards; women generously offering what they have on hand to restore a famished traveler. For wherever Douglas went, from the moment he stepped ashore at the mouth of the Columbia, he was surrounded by families—of the tribes he visited, and of the Hudson's Bay Company employees. His often-used pronoun "they" included the women and children who were part of this larger scene, and these shadowy figures, though seldom explicitly described, played an essential role in his life and work in the Northwest.

In the 1820s, the women in these relationships were all tribal or mixed-blood. Their unions with fur traders, described as *à la façon du pays*, "after the custom of the country," not only secured companionship for lonely bachelors but also expanded trade ties and consolidated political territory on both sides of the equation. Although the marriage of a teenage girl in Philadelphia or Paris at this time was likely to be predicated on the same sorts of factors, it was the cultural stew, the blending of a range of aboriginal tribes with workers across all ranks, that set the fur trade apart. By 1825, generations of "country wives" and mixed-blood children formed an integral part of fur trade society. (The term *métis*, originally meant to describe a person of mixed French and Cree ancestry, has come to indicate any mixed-blood person.)

———

The first fur trader David Douglas and John Scouler met when they disembarked at Fort George was clerk Alexander McKenzie, who had been left behind to decommission the post when the rest of the men headed upstream to build Fort Vancouver. It was McKenzie (no relation to the explorer Sir Alexander Mackenzie) who initially accompanied Douglas and Scouler beyond the ragged stump farm around the post into the coastal rain forest, and explained some of the protocols of fur trade and tribal relationships. Born in Inverness five years before Douglas, McKenzie had been posted on the lower Columbia since 1814, and was married to a daughter of the Chinook headman Comcomly. The fur traders knew McKenzie's wife as "the Princess of Wales," and she was reported to be an influential figure both at Fort George and in her home village.

McKenzie would have communicated with his wife in Chinook jargon, and that also would have been the language he used to introduce his father-in-law to Douglas. In the ensuing months, Comcomly's network of family connections would provide the naturalist with access to cooperative informants who had a deep knowledge of the area's plant life. McKenzie may well have helped to sharpen Douglas's understanding of this local tongue during a subsequent trek north to Willapa Bay and Grays Harbor. But it was the collector's own facility with language that allowed him to put this asset to use in a variety of ways—including the mitigation of a potential conflict between company voyageurs and Upper Chinook people at the Dalles portage the following spring. "Finding two of the principal men who understood the Chenook tongue, with which I am partially acquainted," Douglas wrote, "the little I had I found on this occasion very useful."

Soon after he established his base of operations at Fort Vancouver in the spring of 1825, Douglas met Archibald McDonald, the clerk in charge of accounting at the nascent post. McDonald had emerged from a small village in Scotland to join the Hudson's Bay Company, and had been posted to the Columbia around the time of amalgamation. While working at Fort George, he married another daughter of Comcomly, whom the traders called Princess Raven (several of Comcomly's children carried royal nicknames). In 1824, a son named Ranald was born to the couple, but Raven died soon after. When Archie was reassigned to Fort Vancouver, he left the child in the care of family members at Comcomly's village.

Archibald McDonald

"In company with Mr. Work and McDonald started on a journey by water with a party of twenty-eight men."
—David Douglas

Despite their nine-year age difference, McDonald and Douglas quickly became fast friends, and corresponded for the rest of Douglas's life. Soon after the two men met in 1825, Archie married Jane Klyne, a mixed-blood woman whom he affectionately called "Jenny" and "Madam." He took great pride in teaching her to read and write and bake a Yorkshire pudding. Although Douglas crossed paths with the McDonalds at various trade houses, he made no mention of Jane's hospitality or of her brood of little boys, four of whom were born during Douglas's years in the Northwest.

On the other hand, Archie's son Ranald, who came to live with his father and Jane when they married, fondly recalled the naturalist's presence

Jane Klyne McDonald

"After washing and having a clean shirt handed to me, I sat down to a comfortable dinner."

—David Douglas

during his childhood. Douglas, whose letters to the children of a Scottish friend expressed playfulness and affection, made enough of an impression that young Ranald later described the collector as "my dearly loved friend and companion in the wild woods of the Columbia and Fraser," as if Douglas were a favorite uncle whose appearance signaled fun and adventure. "The Lion in the man awed all," wrote Ranald.

The man usually referred to as the Lion of the Columbia at that time was John McLoughlin, the Hudson's Bay Company's chief factor for the Columbia District and Douglas's official host. During his frequent stopovers at Fort Vancouver, Douglas sometimes lodged in the chief factor's home, where he would have become well acquainted with McLoughlin's wife Marguerite, a mixed-blood woman from eastern Canada whose first husband, Alexander McKay, had served as a valued member of Sir Alexander Mackenzie's 1793 trip across the Rocky Mountains to the Pacific. At Fort Vancouver, Marguerite and John lived with four offspring of their own, as well as Joseph, the teenage son of a Chippewa woman whom McLoughlin had left behind at Red River. This kind of blended family, with connections rippling back to eastern Canada and through tribal bands on both sides of the Rocky Mountains, was not unusual for the era.

During his time in the Northwest, Douglas came to know several traders whose country wives belonged to Plateau culture tribes, and the stories of these families reflect the deeper contact-period history of the region.

Peter Skene Ogden, for instance, was a frequent visitor at Fort Vancouver during Douglas's residence. The son of a Montreal judge,

Ogden had chosen a career as a fur trader. While stationed at Spokane House, he married Julia Rivet, the daughter of a Salish woman and a French Canadian voyageur who had worked for the Corps of Discovery in the Dakotas and then for David Thompson in Montana. Julia, while pregnant and caring for several young children, accompanied Ogden on Snake River expeditions that moved through southern Idaho, eastern Oregon, and northern Nevada.

Finan McDonald, who escorted Douglas on his first excursion down the Willamette Valley, had come to the Columbia District as David Thompson's chief clerk and had married a Kalispel woman called Margaret or Peggy. In 1825, Finan left her and four of their children in the safety of Fort Vancouver while he led a fur expedition southward. David Douglas happened to be present a year later for Finan's reunion with his family, when the separate canoe brigades in which they were traveling met by chance at the Dalles portage. Although Finan had instructions to proceed downstream to Fort Vancouver, he appealed to the agent in charge for permission to change direction and accompany Peggy and the children upstream. "To this request I did not think proper to oppose myself," the agent noted in honor of family devotion. Douglas, who developed a close enough relationship with Finan to trust him to transport a chest of precious plant specimens across the Rocky Mountains, met the family again at Fort Edmonton in

Douglas's Journal, July 10, 1826
"My old Friends Mr. Work and Mr. McDonald handed me my letters."
—David Douglas

spring 1827, and traveled with them in the same canoe brigade as they sped across the Prairies on the Saskatchewan River.

Another trader with whom Douglas became friendly was John Work, a font of knowledge about the natural world. The two men traveled together in canoe brigades and on horse drives, and compared notes on grouse and goats during various encounters. Beginning in 1826, Work would have been accompanied on some of these occasions by Josette Legace, the daughter of an early North West Company voyageur and a Spokane woman. According to an agent working in the district at the time, Work chose to "sign and seal" his alliance with Josette in the presence of company witnesses, and considered himself "safely moored with her for life." When the collector stopped at Fort Colvile in the early spring of 1827, he made no reference to Work's new wife, but was delighted to receive a gift of a nightcap "netted by an Indian girl" from mountain goat fleece. Josette Work, also known as Little Rib, would have been about fifteen years old at the time, and may have noticed that her husband's naturalist friend was going bald and needed a warm nightcap.

The dearth of meaningful detail about women whose contributions were obviously integral to the fur trade has given rise to a school of historians who comb through pay lists, church records, and tribal oral histories to piece together fragments of the life stories of fur trade wives. While their existence can certainly be proved, it is not so simple to understand what kind of world swirled around them. Whether a girl grew up on the inside or the outside of a trade house stockade might represent a huge difference in cultural habits and language skills, not to mention the kind of local plant familiarity that would have been of value to David Douglas. But without close knowledge of how people were raised, there is no way to get a sense of their abilities.

"Look," said an acquaintance after she helped to piece together some mixed-blood family relationships around Fort Vancouver. "You see this social dynamic as the fabric of the fur trade. I see the *métis* that way too. But do you think David Douglas saw it?"

In fact, much of Douglas's work reflects the blurred boundaries between the white, tribal, and mixed-blood worlds. When he wrote down a recipe for salmon on a bed of hazelnuts and huckleberries, or paid for woven goods with European manufactured items, he was moving within mixed cultures. When he settled an argument between voyageurs and tribal

people using Chinook jargon, or added a carefully sounded-out Columbia Salish term for mariposa lily in a scientific paper, he was acknowledging the plastic, ever-changing nature of local speech. When he purchased items at the Fort Vancouver company store such as "6 yards hair ribbon" and "1 fine lady's hat," it seems clear that he must have had his own relationships within that world.

Evergreen huckleberry
Vaccinium ovatum
"Faithful to his proposition, Cockqua brought me a large paper of seeds of Vaccinium ovatum."
—David Douglas

The leader of the 1827 spring express with which David Douglas traveled across Athabasca Pass was Edward Ermatinger, an experienced agent who stood outside the general run of his peers because he had been born in Switzerland and educated in London, and because he was one of the few confirmed bachelors in the entire Columbia District. Around the campfire, Ermatinger entertained Douglas with airs on his flute and artful fiddle playing. He also shared his companion's interest in natural history, stopping at the Arrow Lakes to search for a mysterious "crystal" rock. Edward retired from the fur trade early, but for years afterward engaged

Dr David Douglas
To Fort Vancouver Depot
for the following supplies viz.

			£	
1830 June 28	2 Morocco Colᵈ Belts	1/10	3	8
	2 Pairs Straps w solar buckles	6/	12	..
	1 Grey milled Cap		1	4
July 1	2 Plain Blankets 3 Pts	10/6	1	1
	4 Fine whᵗ Flannel Shirts	7/6	1 10	..
	10 Yds Duck Sheeting	1/6	15	..
	2 ... ditto	1/6	3	..
	1 Pair Round Shoes		11	3
	1 Plain Blanket 3 Pts		10	2
	2 Pairs Long worstᵈ Hose	2/6	5	..
	1 Plain Blanket 3½ Pts		11	3
	6 Lbs Yellow Soap	8	4	..
	2 Plain Blankets 3 Pts	10/2	1	4
	1 Scarlet milled Cap		1	2
	3 Fine striped Cotton Shirts	4/6	13	6
	1 " white Flanᵉˡ ditto		7	6
	1 Pair Long worstᵈ Hose		2	6
	7 Yds Fine Printed Cotton	1/6	10	6
Augᵗ 6	12 Lbs Bar Lead	3	3	.
7	1 Fine Printᵈ Cottᶰ Shirt		6	9
	1 Pair Long worstᵈ Hose		2	6

Columbia Supply List, 1833

Douglas purchased many personal supplies and trade goods from the Fort Vancouver storehouse.

John Work in a lively correspondence that belied his somewhat stiff image in the business.

Douglas had also met Edward's younger brother Francis at Fort Vancouver and Fort Okanagan. Francis Ermatinger had a much looser reputation than the measured Edward, and David "Chalk" Courchane, one of his descendants, isn't sure he deserved it.

"Some people talk bad about Francis," says Courchane, "about his abandoning three wives. But they don't understand the situation.

"Francis Ermatinger's first wife was a Cree woman from the Severn District near Hudson Bay. They had a family till he was posted away to the west. Francis did ask after her, did try to keep in touch, until he heard that she and their children had frozen to death. It turned out they all survived, and the Cree woman remarried another fur agent from the Bird family.

"The next one they called Cleopatra, and I guess she was a handful—sleeping with other men. But she later hung herself anyway, so she wasn't there to be abandoned.

"And Francis did try to keep his third wife, Mary Three Blankets, *Cha-teel-she-nah*. They had a daughter, Three Dresses the Younger; in my family, we call her Grandma Ashley. Mary Three Blankets died when our Grandma Ashley was about six years old, and when Francis came around to look for his young daughter, the family hid her from him. What was he supposed to do?

"Francis wasn't the only one who that happened to, you know. Lots of tribal families would hide their daughters from the fur men. I guess that's why he felt he could start again."

Courchane pauses for a moment. I realize that with those brief glimpses of women associated with Francis Ermatinger, he has tapped into three of the myriad possible outcomes of the mixed-blood relationships that David Douglas glimpsed. During the course of all his travels, the naturalist would have heard stories about casual fur trade acquaintances such as Ermatinger's, and witnessed the dynamic of couples firsthand through closer friends like Archibald McDonald and John Work. No period writer from the Columbia District, scribbling away at fur reports or penning letters back home, would have done more than touch on the complex details of any of these bonds. But then, as now, lots of people would have talked about them at length. It made me wonder what they said about David Douglas.

Courchane is an enrolled member of the Salish-Kootenai tribes of Montana, and his passion is comparing the oral and written histories of his fur trade and tribal ancestors. He has spent plenty of time in the archives, and even more time talking to cousins and friends, listening to them spin their genealogical webs, trying to fit snippets of memory into a coherent pattern of relationships. He sees the interconnections of many of the early families, and traces his own lineage both from a tribal side and from two different fur trade employees.

"I wish I'd paid more attention to my own mother," he says, voicing a common lament. "She told lots of stories that she had heard as a little girl about her grandmother Three Blankets, and some of them had Ermatingers in them. Not as many as the ones about Jaco Finlay and his

Josette Legace Work with children
"We were most cordially welcomed [to Fort Colvile]."
—David Douglas

Indian Tobacco
Nicotiana
quadrivalvis

"I had seen only one plant before, and although I offered him 2 ozs. of manufactured tobacco he would on no consideration part with it."
—David Douglas

people, on our other side. I guess since Jaco and so many of his children all stayed in this part of the country, you naturally hear more about them."

Jaco (pronounced *Jocko*, with many different spellings) Finlay was born in 1768 at the Forks of the Saskatchewan River, the son of a Scottish fur trader and a woman probably of Cree ancestry. Barred from advancement to North West Company partner because of his mixed blood, Jaco worked for that outfit for many years as a clerk, scout, and interpreter, and built Spokane House in 1810. When he passed away at that post in 1828, Finlay had fathered at least twenty-two children by four different wives.

Courchane has chased Jaco Finlay's offspring for some decades now, and has a thick notebook that delineates them across the region and beyond.

"Did you know," he asks me one day, "that David Douglas had a son?"

A mutual friend and dogged researcher had come across a letter concerning a Hudson's Bay Company worker named David Finlay. Identified as "the son of David Douglas," this young man had been killed by Blackfeet on the Flathead River in 1849.

Courchane traces out his line of reasoning. In his May 1826 field journal, David Douglas described two sons of Jaco Finlay guiding him from Kettle Falls through the Colville Valley to meet their father at the abandoned Spokane House buildings, where Finlay was living at the time. Even before he arrived, Douglas had heard stories about Jaco from veteran fur traders. Douglas's friend Finan McDonald, in fact, had toiled with Jaco under David Thompson, and the pair spent some months together at the original Spokane House in 1810 and 11. That was around the time when Finan's Kalispel wife, Peggy, was pregnant with their first child, and Jaco fathered a daughter by his Spokane wife, Teshwintichina. Later Jesuit mission records recorded Teshwintichina's daughter's name as Marie Josephte; oral accounts say her people called her Josette.

That May, David Douglas stayed for only three days at Spokane House, but he did spend the next two weeks collecting in the nearby Colville Valley, where several of Jaco's older children lived. After traveling to the Blue Mountains, the naturalist returned to Spokane House in late July, sharing a salmon breakfast with Jaco's family there. Then he spent another two weeks working around the Colville area before returning to Fort Vancouver. After wintering downstream, Douglas visited Fort Colvile briefly the following spring for a third and final time.

Sticky Currant
Ribes cereum
"Mr. Finlay tells me that the white Ribes in that neighbourhood produces . . . red small solitary berries of a pleasant taste."
—David Douglas

David Finlay is described in the Bay Company's 1849 report as an "Interpreter to the Flatheads." The agent who wrote the original report about this David's fate was a cousin of Archibald McDonald, who would have known all about David Douglas's time in the Columbia country.

"No other David Douglas appears on the Hudson's Bay Company records of the time," insists Courchane. "I've looked, and you know how carefully they kept their pay sheets. And I can't find any David either in oral accounts or in the Jesuit mission records for the first or second generation of Jaco's children. It's not a name you see used by that family."

While Douglas's daybook sheds no light at all on the events implied by this tiny snippet of story, he did record the foods he shared with Jaco Finlay's family. During his very first evening at Spokane House, the naturalist's journal entry consisted mostly of a recipe for preparing a traditional Spokane "moss bread" from the black tree lichen *Bryoria fremontia*.

Bigleaf Lupine
Lupinus polyphyllus
"One of the most magnificent herbaceous plants which have yet come under my notice. In rich alluvial plains."
—David Douglas

He described gathering the lichen and cleaning its black hair of twigs; soaking it until soft; and layering the lichen between grass or leaves on the embers of an earth oven. He watched the mixture cook overnight, then emerge from the oven as a black gelatinous goo that was shaped into cakes before it cooled.

This lichen and several other plants that appear on his daybook list would have been gathered and prepared by women. They include berries, roots, and an aromatic sage growing from rock fissures that emitted "a strong scent like mint." That sounds very much like one of the Spokane tribe's traditional wild teas—both a beautiful flower and, for a man who loved good tea, a welcome pick-me-up.

Courchane ruminates on these data points. He notes that Josette Finlay would have been about sixteen years old the summer that David Douglas spent in the Colville and Spokane Valleys. He recalls that Jaco and his wife had raised other orphaned children during their time west of the Divide. "That's the kind of people they were," he says.

Courchane thinks about the language skills of a boy who would have grown up with a mixed-blood grandfather like Jaco and a tribal grandmother like Teshwintichina, who still made the annual rounds with her familiar relatives: an ideal background for a company interpreter. "If Josette did become pregnant by David Douglas during the course of that summer," he says, "their child would have been, what, about twenty-two years old when he was killed? Doesn't that sound about right?"

"Well," Courchane declares finally, "no one could ever prove it for certain. But who else would it be? To me—to anyone who comes out of this world—there's no kind of surprise here. That's just the way things worked in that time, in that place. David Douglas may have known about his own child, and he may not have. I guess in the long run, all that would have been important for David Finlay is that he was raised by a family who called him their own."

Royal Horticultural Society Medal
Presented to John McLoughlin
"For the assistance rendered to Mr. David Douglas, whilst making his
collections in the countries belonging to the Hudson's Bay Company."

VI.
COMRADES AND MISCREANTS

Bringing the Northwest to London

❧

THE SCHOOLMISTRESSES AND MASTERS of the village of Scone in the early 1800s would hardly have predicted a career as a disciplined scientist for the young David Douglas. Described as "a very spirited boy" with a "taste for rambling," he exhibited a strong preference for fishing and catching birds over the restraints of the classroom. By the time he was ten years old, according to his brother John, he had developed a decided interest in gardening and began spending his summers working on the grounds at Scone Palace, the manor above his home village. The boy obviously had a gift for growing plants, and was bound over as an apprentice during his teen years. Under the tutelage of a knowledgeable master, he took turns in the manor's flower beds, vegetable patches, formal gardens, shrubbery, and arboretum. A coworker later wrote that "no one could be more industrious and anxious to excel than he was—his whole heart and mind being devoted to the attainment of a thorough knowledge of his business." However, the same mischievous nature that had marked his school years occasionally continued to surface. When some of his peers complained to the head gardener of Douglas being a "bad boy," the master tersely retorted: "I like a devil better than a dolt."

SCONE PALACE.

Scone Palace, Perthshire

As an apprentice, David Douglas worked in the arboretum, formal, flower, and kitchen gardens at the manor near his home village.

By the time he turned twenty, Douglas was honing his craft at a larger estate, where he handled the sort of exotic introductions, such as rhododendrons from China and bromeliads from South America, that were quickly gaining in popularity among aristocratic plant enthusiasts. In 1820, his supervisor recommended Douglas for a job at the recently established Glasgow Botanic Garden. He arrived there around the same time as the garden's new director, Dr. William Jackson Hooker, a rising star in British botany. The eight acres on the grounds of Glasgow University already contained over eight thousand species of plants, including an American section, but Hooker was determined to increase their assortment dramatically, and soon Douglas was transplanting new species from far-flung locations that had been sent by Hooker's extensive network of professional and amateur horticulturists.

Meanwhile, Dr. Hooker taught a botany class to beginning medical students, who at that time were all required to learn the identification and basic properties of medicinal plants. Douglas, whose theoretical knowledge of botany was limited to the few texts available at the time, began attending the early morning lectures, which Hooker enhanced with his own highly accomplished illustrations of plant anatomy. The professor "contrived to make botany a first subject, highly interesting and even entertaining, so that his audience laughed as they learned, while regarding themselves as intellectual pioneers." An inspiring figure both inside and outside the classroom, Hooker organized weekend field trips and inaugurated an annual student excursion to the Highlands that welcomed

William Jackson Hooker

"Let me repeat once more how sincerely I am indebted for all I possess to you and the favours and kind attentions I have at all times had from you."
—David Douglas

any local plant enthusiasts who chose to come along. The doctor would load a large tent on a wagon hitched to a small Highland pony, plod into the hills, and set up a base camp at his chosen location. From there, he would lead daily collecting trips of thirty to forty miles through the mountains, then profess to be astonished when, at the end of a session, some of his students "were dreadfully knocked up and ill."

Under Dr. Hooker's tutelage, Douglas flourished as a plant collector, a practical horticulturist, and a technical botanist. In the professor's herbarium he learned to help classify new species according to an emerging natural system of taxonomy that Hooker himself had been instrumental in developing. The student mastered the craft of specimen pressing and artful mounting. He also soon became a welcome guest in the Hooker household, for the professor, at age thirty-four, was a family man with four young children. His home served as both a hub for professional pursuits and a haven of family life: after working with the preeminent botanists in the land, Douglas might find himself shooting game birds with them; between tutorials on freshwater plants, he could acquaint Hooker's two sons, William Jr. and Joseph Dalton, with the basics of trout fishing.

The three years that Douglas spent within this stimulating environment proved to be the perfect preparation for his forays in North America. As a young man, Hooker had himself traveled to Iceland, and before settling down to an academic life had dreamed of "voyaging to the ends of the earth" in pursuit of scientific discoveries. He apparently recognized that same "wandering demon" in his protégé, for when he learned that the Horticultural Society of London was seeking a plant collector in the spring of 1823, he immediately wrote to recommend Douglas. For the rest of his life, Douglas penned long letters to his mentor describing his adventures, and upon his return to England in 1827, he contributed all the knowledge he had gleaned to Hooker's opus on North American plants, *Flora Boreali-Americana*. John Douglas later wrote that "David was invariably guided by the counsels and advice of his first Patron, Dr. Hooker, in every thing that tended to promote the furtherance of science and his future prospects in life."

The Horticultural Society of London had been established two decades earlier by seven influential plant enthusiasts who included such luminaries as John Wedgwood, a partner in his father's famous pottery firm, and Sir Joseph Banks, the aristocratic naturalist who accompanied James Cook's first voyage. The Society's stated purpose was the improvement of horticulture, with the object of collecting "every information respecting the culture and treatment of all Plants and Trees, as well culinary as ornamental," and its roster encompassed a range of social classes and botanical interests. Some of its landed members focused on improving their fruit orchards, while others sought exotic foreign plantings for their formal gardens. A strong contingent of Royal Navy officers pushed for the introduction of new oaks and other tree species that might be of value for shipbuilding. Serious scientific minds debated the virtues of the old Linnaean system versus the newer "natural" method of taxonomic classification.

As with many fraternal societies, such issues evoked strong passions and led to sometimes bitter political infighting over control of the Society's direction.

By 1815, a new generation of leadership emerged at the Horticultural Society when a former inspector general of taxes named Joseph Sabine rose to the position of vice president. A man obsessed with small details, Sabine stabilized a shaky financial situation by enlarging the Society's membership to include aristocrats and contributors outside of Great Britain, and remained in charge of the organization for the next fifteen years. In 1822, Society officers began searching for a suitable site for an extensive garden "in which the new plants acquired might be placed, and their peculiarities correctly remarked." When a nobleman offered to lease thirty-three acres of land in the London suburb of Chiswick, Sabine and his fellows seized the opportunity to inaugurate a garden that would cultivate plants from all over the world.

"Exhibitions Extraordinary in the Horticultural Room"
George Cruickshank's cartoon satirizes a London Horticultural Society meeting.

John Lindley, a young orchid enthusiast the same age as David Douglas, was appointed assistant secretary of the new garden. He soon presented a paper to the Society that stressed the economic importance of new collections, and offered recommendations to facilitate the introduction of healthy plant products to England. Lindley explained the advantages of shipping only established roots and the increased success of bulbs that had been wrapped in paper sacks or packed in dry sand. He outlined methods for ensuring the permanent identification of each collected specimen. For shipping, Lindley emphasized the superiority of wooden boxes over clay pots; the necessity of dove-tailed joints and iron bands to hold containers together through the humidity swings of a long sea voyage; the engagement of a ship's captain in exposing the plants to sunlight and protecting them from cold wind and seawater; and the unfortunate consequences of careless pairings: "Much mischief being done to collections by monkeys and parroquets on board the vessels."

During David Douglas's first collecting expedition, to the mid-Atlantic region of the United States in 1823, he followed Lindley's tenets rigorously. He also made good use of letters of introduction supplied by William Hooker and Joseph Sabine, spending nearly every day in the company of naturalists and horticulturalists who shared his interests. Douglas soaked up the local knowledge of a Dutch peach grower on Long Island, and eagerly listened to a seed propagator at the University of Pennsylvania.

———

The fellows at the London Horticultural Society were well pleased with the results of Douglas's excursion to eastern North America, and when the chance of free passage on a Hudson's Bay Company ship to the Pacific Northwest presented itself in 1824, they instructed him to prepare for a yearlong expedition. Douglas accepted that assignment and much more, discovering such a wealth of productive territory that he made an unauthorized decision to stay in the Northwest for an extra year, and did not reappear in England until the fall of 1827.

As soon as he arrived back in London, Douglas moved into a house owned by the Horticultural Society near Chiswick Gardens and fell to work. He had much to do, beginning with unpacking and sorting the seeds and specimens he had brought with him. He had to check the progress of the material he had shipped home on the *William and Ann* in 1825, which had arrived in England the next spring in time for planting; many had blossomed that summer and fall. John Lindley, who in addition to his duties at Chiswick edited a gardening periodical called *Edward's Botanical Register*, had already featured several of Douglas's Columbia River plants in the magazine, accompanied by full-color illustrations of their blooms drawn from the living plants propagated at Chiswick. Douglas's other collections quickly gained attention as well. Within a month of his return, he had authored a scientific paper that Joseph Sabine read to a meeting of the esteemed Linnean Society. During the next two years, he would write or contribute to more than a dozen other papers, on subjects that varied from currants to horned lizards.

In addition to his plant collections, Douglas was anxious to catalog all the bird and mammal skins he had so painstakingly prepared on the Columbia. To his dismay, the first chest he had sent home on the *William*

St. James Park   Linnean Society   HBC Headquarters

Chiswick   London Horticultural Society   Greenwich

LONDON
1828 – 1829

and Ann had been improperly stored, and the contents were largely ruined. But his later shipments were in fine shape, and Douglas began consulting on their classification with Dr. John Richardson, who was compiling his own material from the first two Franklin expeditions into an opus titled *Fauna Boreali-Americana*. Richardson immediately recognized that these Columbia collections included several intriguing new species, and Douglas's observations on their habits and life histories provided invaluable information for their scientific description. When the first volume of his landmark work was published in 1829, Richardson acknowledged his thanks in the preface: "And here I cannot avoid adding my tribute of praise to Mr. Douglas, for the zeal and intelligence with which he has

London,
1828–1829

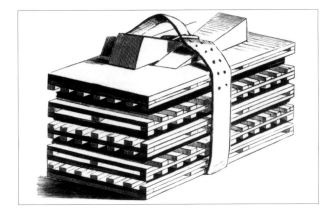

Plant Press

The basic design of the plant press has not changed since Douglas's time.

pursued his scientific researches in North America, and the unusual liberality with which he has communicated his knowledge to the friends of science."

But despite such pleasures of collaboration, Douglas's thoughts began to stray from the work at hand, and his friends began to suspect that he longed to be back in the field. Hooker remarked to a mutual friend that Douglas preferred to tell him in person everything he knew about the plants he had collected rather than write about it himself, adding that he "never seems so unhappy as when he has a pen in his hand."

Douglas's letters to Hooker also began to voice dissatisfaction with his superiors at the Horticultural Society. A few of his complaints, such as the fact that Joseph Sabine never opened one of his specimen boxes from the Columbia, may have been legitimate; others, including his digs at John Lindley's vapid public speaking style, sounded quite petty. Douglas's "quarrelsome tendencies" led to sharp words that shocked some bystanders in the staid Society offices, and the collector wrote to Hooker of his own "outbreakings" as if he were subject to temporary fits of insanity. These incidents leave an impression that Douglas felt insecure inside the complex London world. He was an ambitious man who had accomplished a great deal in a relatively untouched place, and could express that experience with his trademark passion. At the same time, he remained a stonemason's son with a limited formal education, someone who had to worry about his place in a very class-conscious society.

During the same period that the Horticultural Society was prospering under Joseph Sabine, a London publisher named John Murray had found success with a variety of popular titles, as well as literary releases ranging from Jane Austen to Lord Byron. Sensing public thirst for accounts of scientific adventure, Murray published John Franklin's narrative describing his 1819–22 expedition to the North American Arctic within a year of Franklin's return. The account sold very well, and quickly helped finance a second expedition. To some observers, Douglas's trek through the Northwest provided a natural complement to Franklin's saga in the far North.

In 1828, Joseph Sabine took the liberty of discussing such a publication with Murray without consulting Douglas. Although apparently insulted, the collector agreed to take on the project, but pointedly refused Sabine's and Lindley's offers of help to transform his field journal into a proper narrative. Over the next year and a half, Douglas worked on at least two drafts of his Pacific Northwest trip, but the manuscript never quite coalesced into a finished product. Distractions presented real problems for Douglas, and in a letter to Glasgow, he had to admit that he remained "far behind" in his written work. "He has much in his head," Hooker wrote to a friend, "but is totally unfit for authorship."

Douglas's Journal April 27, 1827
Douglas calculated the distances he traveled during his first trip to the Northwest.

1118.

Musk monkeyflower
Mimulus moschatus

"A very beautiful plant; will be a great addition to the garden."

—David Douglas

Yet Douglas appeared to be perfectly fit for a wide range of other pursuits. Joseph Sabine's brother Edward, a retired captain of the Royal Navy, sometimes stayed at the society's lodgings in Chiswick. There he whetted Douglas's interest in geography and geomagnetics, and the two began spending hours in St. James's Park taking readings of magnetic variation. Douglas also began assiduously studying the mathematics required to work out solutions to practical surveying applications.

Meanwhile, the hundreds of plants that he had pressed in the Northwest also demanded his attention. Each specimen needed to be mounted on fresh paper and labeled for placement in the society's herbarium. The many new species he had collected had to be officially described and named. John Lindley helped him sort out around eighteen new kinds of lupines and fifteen penstemons, many of which became popular garden ornamentals. George Bentham, nephew of the economic philosopher Jeremy and a future head of the Horticultural Society, worked with him on other confusing genera, including primroses, monkeyflowers, buckwheats, locoweeds, and milk vetches. In today's world of genetic fingerprinting, the

Linnean Society nomination for membership
Signers included Douglas associates Donald Munro, Robert Brown, Aylmer Lambert, John Lindley, Archibald Menzies, Joseph Sabine, and Thomas Bell.

Clarkia
Clarkia pulchella
"Flowers rose color;
an exceedingly
beautiful plant.
I hope it may grow
in England."
—David Douglas

taxonomy of plant groups is being extensively reshuffled, which makes it all the more impressive that many of the species names originally established by Douglas and his collaborators remain standing today.

Taxonomic nomenclature was then and remains today a contentious realm, full of rivals vying for recognition, but Douglas was usually open to collaboration in such matters. William Jackson Hooker, who was still working in Glasgow, received many of Douglas's specimens, and wrote that his protégé "even begged me to name many of the plants of his collection, that he may be spared the trouble."

Protocol prevents collectors from naming new discoveries after themselves, so it is a testament to the sheer volume of Douglas's work and his standing among his peers that around eighty-five different species of flora and fauna still bear the Latinized tag *douglasii*. The names he chose for his own numerous descriptions better reflect his own thoughts, and it is natural that he referenced William Jackson Hooker, Joseph Sabine, John Lindley, and other scientific cohorts in a variety of ways.

It is striking that of the many dozens of plants that Douglas formally described and named, only two honor a member of the Hudson's Bay Company. Both recognize Nicholas Garry, the deputy governor of their London office, and both represent significant botanical treasures: *Quercus garryana*, now called Garry or Oregon white oak, is the only oak native to the Northwest; Garryaceae, denoting a whole new family of shrubs, is represented on the Northwest coast by the beautiful wavyleaf silktassel, *Garrya elliptica*. All the rest of Douglas's official names acknowledge scientific associates that he met through either Hooker or the Horticultural Society. Of the Columbia District residents who helped him collect the plants that grew there, ranging from John McLoughlin to Cockqua to John Work to Jean Baptiste McKay, not one of their names ever appeared in a plant manual. It is as if Douglas, who by all accounts was a generous man, never quite understood how to connect his informants in the wild Northwest with the very different world of British science.

William Jackson Hooker, looking back at Douglas's many contributions to that field, wrote that "these will constitute a lasting memorial of Mr. Douglas's zeal and abilities; whilst not only in this country, but throughout Europe there is scarcely a spot of ground deserving the name of a Garden, which does not owe many of its most powerful attractions to the living roots and seeds which have been sent by him to the Horticultural Society of London." As a postscript, Hooker added that he had recently heard from a friend who had visited Hammerfest, Norway, "the most northern town in the world, in lat. 71°. The man had spotted one of these Douglas introductions, *Clarkia pulchella* cultivated in pots in the windows of apartments, and very much prized."

Noble fir
Abies procera
"This if introduced would profitably clothe the bleak barren hilly parts of
Scotland, Ireland, and Cumberland, besides increasing the beauty of the country."
—David Douglas

VII.

THE FOREST AND THE TREES

After the Fire

❧

DURING HIS FIRST DAYS AFOOT on the shores of the Columbia, David Douglas wrote that his expectations were fully realized in terms of the varied landscape, the rich soil, and the somewhat intimidating, dense coniferous forests. The only telling exception was the lack of the fine hardwoods that had so enthralled him on his 1823 trip to New York and the Great Lakes. There were no beeches, no magnolias, no locusts, no walnuts, and, as far as he could tell, only one species of oak. That earlier excursion had made Douglas a connoisseur of the wildly diverse genus *Quercus*, and the single example he encountered on the lower Columbia did not at first meet with great enthusiasm: "all the trees which have yet come under my observation are generally low and scrubby . . . in many places, particularly in the upland soils, it dwindles to a mere scrubby bush." Much to his pleasure, however, it turned out that this Northwest oak had many more aspects to reveal.

Garry Oak
Quercus garryana
"Male flowers in pen-
dulous, dense, hairy,
yellow spikes, spring-
ing from the buds
below the leaves."
—David Douglas

As he traveled upstream in John McLoughlin's canoe a few days later, they rounded a bend forty-five miles from the river's mouth and paused at Oak Point, named in 1792 by George Vancouver's lieutenant, William Broughton, in honor of some oak trees spotted there. Douglas laid in leaves from these stately oaks, but the acorns, essential for propagation, were not yet ripe.

At Fort Vancouver he noted that lumber from the distinctive oak was being used for barrel staves and for some of the new buildings being erected at the post. When green, the wood split easily and in regular sections, with little tendency to splinter, but once seasoned it became very hard to work. French Canadian voyageurs told him that the oaks were abundant on hillsides along the river as far upstream as the Cascades of the Columbia.

In early May, on a beautiful plain near the river just a few miles west of Fort Vancouver, he came upon a grove that matched his vision of what an oak should be. In this fertile spot, the species grew straight and handsome to a height of seventy feet, with a girth of six feet at head height. The trees were in bloom, and he collected some of the male flowers, "pendulous, dense, hairy, yellow spikes, 1½ inches in length, springing from the buds below the leaves." The bark of saplings was smooth, but that of older trees was different from any oak that he had ever seen, being "singularly crossed and rugged," of a whitish-gray color, and "divided by regular oblique shallow reticulated fissures."

Late that summer, traveling up the Willamette River, he was delighted to find trees along the banks bearing acorns, "very rare to be had." About the size of large filberts, they had shallow flat cups with soft, silky insides.

During a longer excursion through the Willamette Valley in 1826, he saw additional examples of the "beautiful solitary oaks, scattered here and there." They provided the perfect refuge, as Douglas witnessed on one memorable occasion, for a hunter who was being chased by a grizzly bear.

Back in London in 1828, Douglas revisited a monograph he had written on North American oaks based on the information he had acquired during his trip to the eastern seaboard. This manuscript, never published during Douglas's lifetime, illuminates both his developing collection methods and his aims as a budding scientist.

Douglas was a voracious reader, and he certainly hoped to someday join the ranks of respected collecting botanists who had preceded him. For this initial effort, he used the published works of Frederick Pursh and André Michaux as his guides. The French Michaux had been almost forty years old when he arrived in New York in 1785 and caught Thomas Jefferson's eye as a man who could help broaden a fledgling nation's scientific

Mt. Washington, Willamette Country
"Country undulating; soil rich, light, with beautiful solitary oaks and pines interspersed through it."
—David Douglas

Garry Oak
Quercus garryana
"Acorns, sessile in pairs, about the size of a large filbert, with a shallow flat cup, soft and silky on the outside."
—David Douglas

horizons. Michaux established a garden in the wilds of New Jersey to hold his collections, then traveled to Philadelphia to meet famed Southeastern collector William Bartram. In 1792 he met with fur traders in lower Canada and, guided by a mixed-blood interpreter, attempted a canoe expedition that stalled halfway to James Bay. Although Michaux's and Jefferson's master plan for an expedition to the Pacific in the early 1790s was squelched by political concerns, he carried out more than a decade of work in the New World, and produced several books, including, in 1801, *The Oaks of North America.*

Frederick Pursh, a native of Germany, had begun his career as an assistant to Dr. Benjamin Barton at the University of Pennsylvania in 1805, and inherited the task of describing Lewis and Clark's plant collections after Barton failed to complete the assignment. In 1814, Pursh published *Flora Americae Septentrionalis*, which included one hundred thirty-two species from the Corps of Discovery. Douglas absorbed the entire volume, especially all the genera of trees. "Of the thirty-four species [of oaks] enumerated by Pursh as natives of the vast continent of North America," he wrote in the introduction to his paper, "I was so fortunate as to meet with no less than nineteen." He treated each of these nineteen species as a small obsession, interviewing all manner of woodworkers and judging field characteristics over the breadth of every range. In the course of writing his manuscript, Douglas paid respectful homage to his predecessors, but did not shy away from the value of his personal observations.

Because Douglas never published his oaks paper, he had the freedom to amend it after he returned to England from his first sojourn in the Columbia country. In its surviving form, the treatise's last entry covers *Quercus garryana*—named by the collector for Hudson's Bay Company deputy governor Nicholas Garry "as a sincere though simple token of regard." Douglas waxed enthusiastic over the tree we now call Oregon white oak or Garry oak, allotting it more space than any other species in the monograph save for his initial eastern white oak. He admired *garryana*'s handsome straight form, its peculiar bark, and the downy twigs of young saplings. He thought the tough texture and superior durability of its wood would make Garry oak a valuable general-purpose tree for the timber industry, and especially for shipbuilding.

Douglas observed that the Pacific Northwest's oak was common on floodplains with good clay soil within two hundred miles of the Pacific, and especially so on the north banks of streams within sixty miles or so of the sea. Garry's oak "does not form thick woods as is the case with the Pine tribe," Douglas commented, "but is interspersed over the country in an open manner, forming belts or clumps along the tributaries of the larger streams." With an eye for the practical, he added that such rivulets would be very convenient for floating oak logs downstream for processing.

Biologist Peter Dunwiddie isn't as interested in saw logs as he is in restoring the Garry oak communities that David Douglas described in his species account—those belts and clumps of open oak woods interspersed through both wet savannahs and drier prairie habitats west of the Cascade Range and up the Columbia Gorge. Dunwiddie, along with other Garry oak enthusiasts from the Willamette Valley to British Columbia's Gulf Islands, would like to experience that look again.

The attraction of open Garry oak woodlands can be glimpsed from simple visions: the array of camas, lupine, spring gold, and western buttercups that bloom each spring on neglected land up and down Interstate Highway 5; the crowlike flaps of a Lewis's woodpecker eating up open space between two sheltering oak crowns on a ridge just east of the

Garry Oaks in Columbia Hills with Lupine and Balsamroot
"Interspersed over the country in an open manner, forming belts or clumps."
—David Douglas

Columbia Gorge; the distinctive quick barks of western gray squirrels, which have lost ground to introduced squirrels as the range of their home hardwood shrinks; the flit of the skipper butterfly known as the western oak or Propertius duskywing (*Erynnis propertius*), which can link its whole existence to a single isolated tree.

Today most of the west side's open oak woodlands have been converted to agriculture or swallowed by development, but even the remnant patches do not answer to the descriptions penned by Douglas. Older Douglas-firs seem to overgrow the oaks and constrict their crowns, while crowding by younger coniferous saplings reduces the hardwood foliage and causes noticeable die-back among branches. Low shrubs such as snowberry and hairy honeysuckle blanket understory that used to be open space, and greatly reduce the diversity of wildflowers that Douglas's collections show were once part of the scene. The soils test out as thin and unproductive. "Understanding which ecological processes maintained these oak stands in the past, and how they have changed in recent years, is critical if they are to be protected and managed as viable communities," writes Dunwiddie.

A wide range of studies on Garry oak stands throughout the region point to a complex of reasons for the infill of Douglas-fir. Climate change may have altered moisture regimes. Livestock grazing came in with white settlers in the mid-nineteenth century, followed by extensive rounds of logging. But the single most important factor appears to be the way Native Americans throughout the region systematically set fire to these open oak woodlands "to improve hunting conditions and to promote the growth of plants used for food, fiber, and medicinal purposes." For more than two decades, Dunwiddie has served as a fire boss on various preserved patches of oak, trying to emulate those prehistoric burns, and he appreciates the subtleties of tribal management. "Every fire is different," Dunwiddie says, underscoring the myriad variables that lead to an intuitive feel for when and where to set off a blaze.

On one study plot, Dunwiddie and his associates aged a mixed stand that included both oaks and Douglas-firs over four centuries old. Archaeological data showed burn patterns that averaged every seven years until the middle 1800s, then disappeared almost entirely with the advent of white settlement and the decline of native populations. From then on, the ratio of Douglas-fir to Garry oak rose in distinct pulses as the forest increased its historic density by a factor of ten.

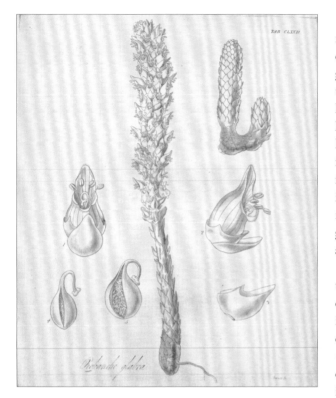

TAB CLXVII

Orobanche ludoviciana

Ground-cone
Boschniakia
hookeri
David Douglas
collected several
members of the
broomrape family in
open oak woodlands
near Fort Vancouver
in 1825.

Douglas saw the effects of native-set fires in oak woodlands during his journey with a fur brigade up the Willamette River in the fall of 1826. When he asked local Kalapuya people what they had in mind with such burning, he received answers that ranged from opening up the underbrush for better deer hunting to clearing an area "in order that they might better find wild honey and grasshoppers, which both serve as articles of winter food."

More subtly, Douglas observed the small annual plants that quickly returned after set fires in Garry oak communities. He collected a broomrape he called *Orobanche ludoviciana* (Louisiana or Suksdorf's broomrape) and identified its habitat and habits. "Found on the alluvial plains of the Columbia; parasitic on the roots of various grasses which have been burned by the natives in the autumn for the purpose of affording a tender herbage in spring for their horses; abundant." He also found the beautiful small-flowered deervetch (*Lotus micranthus*) to be "common on soils where wood has been destroyed by fire, on the shores of the Columbia."

In monitoring burns through oak woodlands, Peter Dunwiddie has witnessed the same kinds of effects. He noticed that after one successful controlled burn, the annuals sprang up in abundance for two or three years, then declined. He observed similar patterns, but to a lesser degree, in many of the other sites he has burned. The reason, he believes, is that most of the native annuals have disappeared from the seedbank of those more populated sites, allowing nonnative grasses and invasive weeds such as Scotch broom to fill the post-burn niche. The native deervetch, first collected by David Douglas, is one of the few annuals that still occurs pretty frequently, and which Dunwiddie can still watch bounce back strongly after a fire.

Such burning regimes do not offer a black-and-white formula for restoring a landscape. On one site, Dunwiddie set his first fire in 1987, then repeated the process in 1996; in their aftermath he recorded reduced cover of the native grasses he was trying to encourage, but an increase in camas, a food root highly desirable to all the tribes. Several oral accounts describe camas bounty as a primary benefit of such fires.

Dunwiddie points out that the tribes before contact had dozens of generations to develop the methods and soil tilth that suited their needs. Today their descendants emphasize that each copse of oak, each expanse of prairie or savannah, is unique. Pulses of tiny plants, such as the deer-vetch and a cluster of broomrape here and there, offer hints at the different worlds David Douglas walked through, with Garry oak "interspersed over the country in an open manner." While we may never be able to re-create the ecosystem that Douglas saw, Dunwiddie's sites—some now a quarter century back into a burning regime—can offer echoes that include a reconstituted suite of flowering plants.

When Douglas arrived back in England from his first Columbia trip at the end of 1827, he was on fire with details of the region's coniferous forests. In his first scientific paper, devoted to the wondrous sugar pine, Douglas delineated the tree's range in a far wider arc than he had covered on his own, assembling information he presumably gained by questioning Umpqua tribal members and employees of the Hudson's Bay Company. He pinpointed ridges of sandy soil east of the Coast Range as the setting where the pine attained its greatest size, and noted that it grew only as scattered individuals among pure stands of other more common conifers. He proceeded through a minute description of every aspect of the tree's anatomy, in fluid prose decorated throughout with the Latinate vocabulary of his profession. The needles, which came in bundles of five to a sheaf, he found to be "rigid, of a bright-green colour, but not glossy, and from minute denticulations of the margin are scabrous to the touch." He did not fail to remark on the change from pitch to sugar brought on by systematic burning, and made sure to include the nuts as an additional resource:

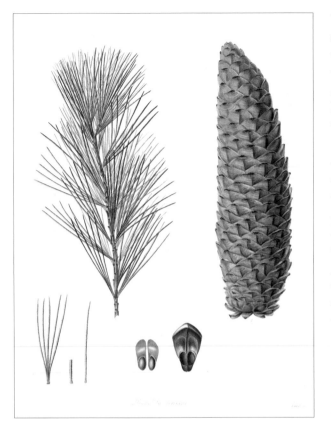

Sugar pine
Pinus lambertiana

"At midday I reached my long-wished-for Pines, and lost no time in examining them and endeavouring to collect specimens and seeds. New and strange things seldom fail to make strong impressions."
—David Douglas

"The seeds are eaten roasted, or are pounded into coarse cakes for their winter store." He included the Umpqua word for the tree, *Nat-cleh*, culled from a language he struggled with throughout his visit there.

As was his habit in all his writings, Douglas added the experiences of well-known predecessors, citing Captain Vancouver's coordinates of Port Orford as the western limit of the sugar pine's growth, and recalling a dessert of large pine nuts that Archibald Menzies enjoyed at a California mission as a possible clue to the southern extent of its range. Douglas nodded to Aylmer Bourke Lambert, Esquire, the vice president of the Linnean Society, "whose splendid labours in investigating the genus *Pinus* are too generally known and appreciated to require any eulogium from me." Then Douglas provided the ultimate eulogium by formally naming his prized tree *Pinus lambertiana*. Lambert repaid him in kind when he published volume two of his *Genus Pinus*. The book featured sumptuous illustrations of Douglas's cone collections, including the enormous sugar pine, the strikingly colored Pacific silver fir, and the porcupine-quilled bristlecone fir—all forever seared into the memory of tree enthusiasts as springing from the hands of the stonemason's son.

———

When Douglas awoke on his first Columbia River morning on board the *William and Ann* in the pouring rain, he seemed somewhat uneasy in the coastal habitat, those "dense gloomy forests, two-thirds of which were composed" of the gigantic odd species that was soon to bear his name. During his first trip to California he felt a similar wave of gloom, declaring that redwood groves north of Monterey Bay "give to the mountain a peculiar—I was going to say an *awful*—appearance, something that tells us we are not in Europe."

No matter how hemmed in he felt in the coastal forests, Douglas always had a naturalist's ability to delight in any new species, and within a few weeks of his arrival he was extracting every nuance from the variety of the trees that grew there. By his second summer in the Columbia, he had a variety of bark and trunk samples scattered around his lodgings, and paper envelopes filled with seeds from the species that appeared to hold the most promise as ornamentals and commercial timber in Great Britain.

He was not alone in his interests. Over the next few years, Douglas's scientific collections ran parallel to the development of timber products as part of the Hudson's Bay Company's regular scheme of trade. As the *William and Ann* prepared to sail back to England in the fall of 1825, two cedar planks were loaded on board, bound for inspection by Her Majesty's craftsmen in London, who would study their qualities as worked wood. The following summer, a pair of twenty-foot Douglas-fir planks were shipped back for the same purpose.

In order for Pacific Northwest tree species to be grown in Great Britain, nurserymen there would require more detailed information about their range, abundance, soil requirements, growth habits, and potential. In 1827, only one man possessed that knowledge, and during his twenty-four-month pause between trips to western North America, David Douglas produced a monograph called "Some American Pines," that is similar in style to his oak manuscript.

In the early 1800s, the pine family encompassed all the conifers we now call fir, spruce, hemlock, and larch. Douglas described many of these with the same enthusiasm he had applied to the sugar pine, emphasizing the qualities of their woods, and conjecturing on which species might perform best in the latitudes of his home hemisphere. Along with his excitement at discovering new species in their native habitat, he often seemed to be peering into the future of silviculture.

Douglas began his paper with a tree that had been first described by his mentor Archibald Menzies, and that had been recently sprouted at Chiswick from seed supplied by Douglas. Hooker had already named the species

Douglas-fir
Pseudotsuga
menziesii

"One of the most striking and truly graceful objects of Nature."

—David Douglas

Douglas pine, *Pinus douglasii*, but the author made no mention of that honor as he tore through his detailed account. The sheer size of this tree, of course, was bound to astonish any European. "Behind old Fort George," Douglas noted, "there stands a stump of this species which measures in circumference 48 feet, 3 feet above the ground, without its bark. The tree was burned down to give place to a more useful vegetable, namely potatoes." He laid out the remarkable difference in form between gigantic solitary Douglas-firs growing on rugged upland soils, of classic pyramidal shape with large lower branches, and tightly grown forest stands of self-pruning spars limbless to upwards of a hundred and forty feet, "which frequently attain a greater height and do not exceed even 18 inches in diameter close to the ground." Once again showing his critical knowledge of the literature, he wondered about a tree that Lewis and Clark had mentioned in their journal and estimated to be at least three hundred feet tall; the most accurate measurement he had come up with was two hundred twenty-seven feet, which seemed impressive enough.

Drawing on a variety of sources, Douglas delineated an extensive range for this species. He marveled at its ability to tolerate the large variation of moisture, soil, and temperature conditions between coastal and interior habitats, and correctly predicted that in time its boundaries would be extended even farther than his initial description. He found its wood "straight and regular in the grain, fine, heavy, and easily split; the layers or rings of a darker tint, closely resembling the timber of the well-known Larch," and thought that it "may be found very useful for a variety of domestic purposes; the young slender ones exceedingly well adapted for making ladders and scaffold poles, not being liable to cast; the larger timber for more important purposes." His single reservation was that "whether it will prove durable or not remains yet to be known." In Douglas's time, lumber represented only one aspect of a tree's value, so he also made sure to consider the properties of its resin for naval stores and its density for charcoal. "The coal is moderately hard, bulky, brown, which might be expected from the great quantity of gaseous matter it contains."

In summing up, Douglas clearly understood that this peculiar tree could well be one of his most valuable collections. "Being an inhabitant of a country nearly in the same parallel of Latitude with Great Britain . . . gives us every reason to hope that it is in every respect well calculated to endure our climate and it will prove a beautiful acquisition to English Sylva if not an important addition to the number of useful timbers."

With his own eponymous pine as a benchmark (it took decades for taxonomists to settle on its current Latin name, *Pseudotsuga menziesii*, and the common name of Douglas-fir), he proceeded to work his way through the region's most significant conifers, always displaying a knack for capturing some salient point of growth, wood quality, or prospective benefit. He thought Sitka spruce "may become of equal if not greater importance" than Douglas-fir as a timber resource, "as it possesses one great advantage over that one by growing to a very large size on the Northern declivities of the mountains in apparently poor, thin, damp soils; and even in rocky places . . . This unquestionably has great claims on our consideration as it would thrive in such places in Britain to become a useful and large tree." He continued with similar assessments of noble, Pacific silver, and grand fir, Western larch and hemlock, and lodgepole pine, giving each one the attention it deserved.

The open woodlands of the Columbia's dry Interior have a very different look and feel than the coastal rain forests, and a reader of Douglas's journal can sense a lift in his spirits upon his arrival at the mouth of the Spokane River in early April 1826. "This part of the Columbia is by far the most beautiful that I have seen," he wrote. "Very varied, extensive plains, with groups of pine-trees, like an English lawn." The following month, as he rode through old-growth stands of ponderosa pine between the Colville and Spokane River drainages—open forests managed for bunchgrass, berry-producing shrubs, and root plants by tribal burning—he was equally

Ponderosa pine
Pinus ponderosa

"Seeds oval, black, encircled by the base of the wing, the inner side of the seed being nearly covered."
—David Douglas

impressed. "Delightful undulating country . . . The scenery is picturesque in the extreme . . . At seven in the morning gained the summit of the last range of hills between the two rivers, and had one of the most sublime views I ever beheld."

When he addressed those well-spaced ponderosa pines in his writings, Douglas expressed concern that "the young trees are liable to injury by a singular species of mistletoe which grows so rapidly in such abundance that it in time completely destroys it." He explained how Plateau people used this pine: "The seeds of this are eaten by several of the native tribes raw but more generally dried or roasted in the embers. The natives also strip the bark for cambium." And despite the mistletoe problem, he remained optimistic about the commercial possibilities of ponderosa pine, whose wood he declared to be "remarkably clean-grained, though somewhat coarse in texture, smooth, heavy, reddish, works fine." He seemed even more intrigued by the fact that the wood was "impregnated with a copious resin . . . a great portion of turpentine could be extracted."

It's May, and the morning air lilts with a vanilla tang of pine resin. On an open slope I squat before a knee-high mound of thatch ants, watching its inhabitants open up their dwelling for the day. Among the mosaic of redheaded workers, vent holes begin to appear, breathing air far underground to the egg-laying queen. Scouts radiate outward from the mound, circling my feet as they sense their way through a universe defined by huge old ponderosas, well spaced. The orange-red bark on all the trees flakes into thick jigsaw pieces perfect for an ant to climb up or a nuthatch to clamber down. Old snags shed thigh-sized limbs from high above. Yellow bells and spring beauties, two prime roots for tribal digging, snake through the bunch-grass in untold thousands. A lot of spring bird noise floats through the air—chickadees, kinglets, thrushes, flycatchers, warblers, vireos, tanagers, and more—but all of it sounds distinct and melodic. The proportions within this grove of old growth, creeping up the slope to lip onto a flat in the Little Pend Oreille National Wildlife Refuge, feel just right.

Ponderosa pine
"Trees tall, straight, very elegant, carrying their thickness to a great height. The bark is very smooth, tawny-red. Fruit perfect in September of the second year."
—David Douglas

There is no way to know whether David Douglas climbed this rise during his several weeks of wandering through the Colville Valley in 1826, but he did swim the Little Pend Oreille River a few miles downstream at least three times, and tried to dry out while camping "a few yards from the river, in the shade of some pines." Many modern observers believe that he would have seen a lot of ponderosa stands like this one, with trunks twelve to twenty-four feet in circumference, certain branches bristling with mistletoe, white-headed woodpeckers dancing down their flanks, and flammulated owls nestled silently in hidden cavities. As the season wore on, the succession of wildflowers would have included rein orchids, clarkias, and mariposa lilies, scattered in the sunlight on exposures that suited them.

Douglas saw each of those plants when he rode north through the Colville Valley that August, but lost all the seeds he had collected from them when his horse mired down while fording the next major creek to the

Showy penstemon
Penstemon speciosus

"Near Oakanagan and on the Spokan River. A splendid species, with very brilliant blue flowers slightly tinged with blush."

—William J. Hooker

north. The ensuing mud battle left him with a knock on the head and thoroughly soaked his field note-books, but such mishaps were to be expected by someone who worked in the field. In time he sent back good seeds and specimens from the forest habitats of this valley, which were used both by Horticultural Society gardeners at Chiswick and by William Jackson Hooker in his written publications.

The climate in the British Isles proved far too wet for ponderosa pine, but other tree species Douglas collected while in the Columbia country and California found more favor on his home turf, and have been widely planted there for almost two centuries now. Stately clumps of Douglas-fir in Scotland, some propagated from cones he sent home in 1825, reinforce Douglas's status as a botanical hero. Many British silviculturists regard him as a pioneer, pointing to commercial stands of Sitka spruce in Ireland as examples of his vision. Yet, as historical botanist Oliver Rackham of Cambridge University reminds us, a dynamic woodland is impossible to create from scratch. Rackham thinks that among Douglas's introductions to Great Britain, the timber trees "tend to be fast-growing, expensive, and of poorer quality than when grown in their homelands." He acknowledges the success of Douglas-fir, and understands the appeal of Monterey pine and grand fir as ornamentals. But for Rackham the western red cedar has never made much of a commercial mark, and current climate tendencies may spell trouble for Sitka spruce. Monterey pine, which Douglas sent home from California, has not done well as a commercial tree either, and Rackham considers lodgepole pine almost useless except as a tax shelter, noting that it has been banned in New Zealand as an invasive. "What is missing," he emphasizes, "are the ethnography and the ancient trees that were left behind in their homelands."

That is exactly the kind of context modern foresters are trying to recapture as they manage tracts of native trees for more complexity. Given that no one can know what the oak prairies of the Puget Trough looked like when Douglas tromped up the Cowlitz River in 1830, or how one south-facing slope along the Little Pend Oreille has changed since the summer of 1826, that seems like a daunting task. But several of the old ponderosas in this grove must have been here when he visited, suffusing the air with the scent of running sap. Certainly Douglas would have stepped around a few thatch ant mounds during his wanderings, and listened to the three-noted call of a western tanager, repeated over and over from on high. Such single threads, however tenuous, offer the promise of weft and warp.

Agama Douglasii.

Short-horned lizard
"This beautiful and highly interesting species was found by Mr. David Douglas."
—Thomas Bell

VIII.

THE WISE ECONOMY OF NATURE

Adapting to the Landscape

❧

DURING THE EARLY NINETEENTH CENTURY, public and private museums, scientific organizations, and hobbyists strove to fill "Cabinets of Natural History," and traveling naturalists were urged to send home specimens of as many different life-forms as possible. As one prominent zoologist of the time explained, "The economy of animals can only be studied when the functions of life are in full activity; their haunts must be explored, their operations watched, and their peculiarities observed in the open air. But in order to acquire a more accurate knowledge of their external form, and to investigate their internal structure, it is absolutely necessary to examine them in a dead state."

Prairie Wolf (Coyote)

"Mr. Work had a solitary skin of the small wolf of the plains, a singular variety and curious from its being the deity of the Flathead tribe of Indians."
—David Douglas

Joseph Sabine, secretary of the London Horticultural Society, shared this fascination, and expressly instructed Douglas to collect not only plants, but also "other objects of natural history" during his sojourn in the Northwest. This assignment would have created no hardship for Douglas, for although his career was focused on the plant kingdom, he never lost his boyhood interest in the natural world at large. His childhood talent for capturing live birds and mice would prove to be excellent practice for the observation and patience that promised success as a collector, as did his self-avowed prowess as a crack shot.

On the long sea voyage from London to the Columbia, he had ample opportunity to practice his dissecting skills in company with John Scouler, the ship's surgeon, who had a special interest in comparative anatomy. Although his first month on the Columbia passed in a flurry of plant pressing and seed gathering, Douglas also devoted time to his secondary assignment. Returning to Fort Vancouver in late May from a short trip upriver, he noted that he had "increased my collection of plants by seventy-five species, and also killed four quadrupeds and a few birds."

During an excursion up the Willamette River that August, he brought back the skins of a young white-tailed buck, some curious mice and dormice, a coyote or a small wolf, and a rodent that he was certain would form a new genus. For his preparations of these specimens, he carried a dissecting kit that would have included knives, scissors, needles, silk thread, arseniated soap, preservative powder, and cotton stuffing. After dispatching his subject, his first task was to carefully measure the body and its limbs and record detailed notes on its teeth, claws, coloration, and other distinguishing features. Then he could begin the tedious process of skinning the carcass, cleaning the flesh and muscle from the skull, and scooping out the brain tissue. Once the skin was free, he brushed the entire inner surface, the skull, and any remaining bones with a lather of arseniated soap (carried in a small tin box labeled "POISON"). This procedure required great care, for the tiniest

bit of soap caught beneath a fingernail would eat away the flesh. After filling any incisions with preservative powder, he stuffed the body cavity with cotton, then stitched the skin back together. When the *William and Ann* sailed for London from the Columbia River in the fall of 1825, her cargo included a large chest layered with a selection of bird and mammal specimens.

During the winter months around Fort Vancouver, with little to be done in the field of botany, Douglas devoted a great deal of time and attention to the local fauna. He compiled an informal list of bird and animal species, including interesting behavior he had witnessed, as well as life history and range details supplied by all manner of local sources. He wrote accounts of many Columbia birds, paying particular attention to the wonderful variety of grouse he had seen and heard about—as an avid sport hunter, wing shots were a source of great pride. He prepared specimens of magpies, eagles, crows, hawks, geese, and one female swan. He stayed up for six nights waiting for a chance to bag a large horned owl by the light of the moon, and tried repeatedly to bring down one of the California

California Vulture (Condor)

"Except after eating, they are so excessively wary, that the hunter can scarcely ever approach sufficiently near, even for buckshot to take effect on them."
—David Douglas

Bobcat pelt

"Mr. Douglas brought a specimen of a Lynx from the Columbia River."
—John Richardson

condors that wintered on the river. "I killed only one of this very interesting bird with buckshot," he wrote, "one of which passed through the head, which rendered it unfit for preserving; I regret it exceedingly, for I am confident it is not yet described." After that unfortunate blast, he asked every hunter he met to keep an eye out for the outsized buzzards, and promised to reward anyone who could deliver an intact carcass.

In contrast to the abundance of bird life, Douglas found the mammals along the lower river somewhat of a disappointment, lamenting that "the variety of species of quadrupeds is not I think so great as in many other parts of America." Like other armchair tourists of the American West, he probably had in mind the herds of big game that Lewis and Clark had witnessed on the Great Plains. In keeping with his character, however, he set about learning everything possible from the situation at hand. Puzzling over the differences between the elk and two types of deer he encountered, he gave taxidermy lessons to several company hunters and supplied them with preserving powder in hopes of obtaining specimens for comparison. He watched tribal hunters call in bucks with flutes fashioned from sections of cow parsnip. While hunting on horseback with a companion and two dogs in February, "we raised a large Lynx . . . a small bull and a terrier dog immediately seized the lynx by the throat and killed it without much trouble. It was a full-grown female." Douglas skinned the cat and added its pelt to his growing collection. John Richardson later helped him identify the cat as *Felis rufus*, the Canadian lynx's close relative that today we call a bobcat.

Douglas initially saw some of his most interesting mammal specimens as trade skins or tribal clothing. In the Willamette Valley, he marveled at the array of colors on a fox robe worn by a Kalapuya child, and tried to purchase either the robe or some of the fox skins, but "too great value was put on them." He had better luck in acquiring a robe made from the pelts of the mountain beaver, known among the coastal tribes as *sewelel*.

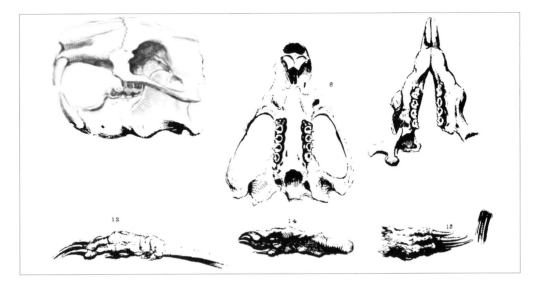

The robe contains twenty-seven skins, which have been selected when the fur was in prime order. In all of them the long hairs are so numerous as to hide the wool or down at their roots, and their points have a very high lustre.

Douglas tried on several occasions to capture one of the elusive *sewelels*, and eventually the Chinook headman Cockqua obliged him with several specimens of males and females of different ages. Dr. John Richardson used these as the basis for the first accurate description and illustrations for a new genus he christened *Aplodontia*.

Douglas procured a wealth of information about other members of the rodent tribe, perhaps because in the course of his plant collecting duties he often found himself competing with them for ripe seeds and roots. On a memorable night at Fort Walla Walla, he was able to apprehend a bushy-tailed woodrat in the act of devouring seed packets, dried plants, and a shaving brush. In the Willamette Valley, he snared an "animal known on the banks of the Columbia by the name of the Camas-rat, because the bulbous roots of the Quamash or Camas plant form its favorite food." At Fort Vancouver, he spent time observing the antics of a similar animal that he believed to be a different species, which he called the Columbia Sand-Rat, with large cheek pouches that resembled the thumb of a lady's glove:

Aplodontia
"The legs are very short, and are covered down to the wrists and heels with fur similar to that on the body. A little above the wrist joint, on the inner side, there is a small tuft of stiff white hairs."
—John Richardson

These little rats are numerous in the neighborhood of Fort
Vancouver, where they inhabit the declivities of low hills, and burrow in the sandy soil. They feed on acorns, nuts, and grass, and
commit great havoc in the potatoe-fields adjoining to the Fort, not
only by eating the potatoes on the spot, but by carrying off large
quantities of them in their pouches.

He kept an eye out to see what happened next: "When in the act of
emptying its pouches, the animal sits on its hams like a marmot or squirrel, and squeezes his sacks against the breast with his chin and fore-paws."
Interested in those cheek pouches, he captured a female in her nest, along
with three of her young, and noted that "the outside of the pouches was
cold to the touch, even when the animal was alive." As he dissected the
animals, he reported that the insides of the pouches "were lined with
small, orbicular, indurated glands, more numerous near the opening in
the mouth." When John Richardson examined the specimens in London,
he believed that Douglas had discovered a new species, which he named
Geomys douglasii as "a small tribute of respect for the zeal and intelligence
of its active and diligent discoverer." Modern taxonomy, however, has
joined both the camas-rat and the sand-rat to the species of the mazama
pocket gopher, *Thomomys mazama*, choosing to honor the Crater Lake
volcano that spread a thick layer of black ash across the region. For any
gophers that survived that blast, the shift from camas bulbs to Irish potatoes must have paled in comparison, and Douglas's view of their double
life only presaged other logical adaptations to the green garden plants,
roots, and seeds of human endeavors.

———

Although Douglas's rodent collections and rough notes supply only fragments of the animals' larger life histories, they do provide clues as to how
the habitats and behavior of certain species have changed over the past
two centuries. Each spring, then and now, as the snow wears off hillsides
throughout the Northwest, ropes of wet soil slither across the slopes like
cryptic runes. The dark strands represent pocket gopher tunnels, pushed
up as the rodents work through their busy winter routines on the horizon
between snow and bare ground. These gophers, common from alluvial

soils along every river's edge clear up to stony meadows in the high country, never seem to stop digging.

There are more than a dozen species of pocket gophers spread across our continent. They live in North America and nowhere else. All look very similar, and share the same compulsive habits. Their fossil record stretches back to the Oligocene, thirty million years ago and more; the pocket gophers that burrow through the Pacific Northwest today are not very different from their ancestors who shared space with saber-toothed cats, New World camels, and mastadons.

As the ropes of winter's plowed-up earth subside into the ground, mounds of freshly dug soil identify the entry point for spring's active tunnels. Each mound usually shows a distinctive fan shape, growing out from a dirt plug that seals in the darkness of the gopher's world. Although you can occasionally find one of these holes open, or catch sight of a gopher foraging on plants topside, the vast majority of their time is spent below the surface.

Pocket gophers are easy to identify because the skin around their lips extends behind their front teeth, leaving big yellow incisors exposed at

Pocket Gopher
"Mr. David Douglas informs me the animal is known on the banks of the Columbia by the name of the Camas-rat, because the bulbous root of the Quamash or Camas plant forms its favourite food."
—John Richardson

all times. The large cheek pouches that inspired the animal's common name drop as vertical slits from each corner of the mouth, and lead to fur-lined, reversible openings tailored like two fine purses. Their tiny ears and eyes, combined with oversized, recurved front claws, make pocket gophers beautifully adapted for tunnel life, where they can grab succulent young plants by the roots and pull them down into their burrows, or fill their pockets with plump and crispy bulbs.

In the wild, the pocket gopher's lifestyle plays a key role in the creation of healthy tilth—churning subsoils to the surface, aerating compact ground, and aiding in moisture conservation. According to David Douglas, fur trade voyageurs called them *gauffres*, which can be translated as "honeycombs" or, better yet, "waffles." Any early spring walker through a shrub-steppe swale in the Columbia Basin will tread on the endless waffle pattern of overlapped gopher diggings.

Ever since the arrival of humans in their world, pocket gophers have competed with the newcomers for various roots, greenery, and seed resources. This situation seems to have grown more testy over the past couple of centuries, and in today's Northwest *gauffres* are seen as serious threats to everything from commercial hay crops to backyard gardens to commercial timber crops.

In the 1950s, tree growers in eastern Washington realized that their clearcut plantings of young ponderosa pine trees were dying off due to pocket gophers chewing on their tender roots. A survey revealed that the same thing was happening in reforestation plots all over the West, and intensive poisoning campaigns, usually employing strychnine, were implemented with all manner of contraptions to deliver a killing dose. None of them worked very well, and a research biologist concluded that the practice of replanting clearcuts with a single tree species was probably encouraging the explosion of gopher activity. As the camas-gathering tribes on the Columbia could have told the foresters, this is an animal that will teach you how to share.

———

While visiting Philadelphia in 1823, Douglas stopped by Charles Wilson Peale's museum to view a display of curiosities collected by the Lewis and Clark Expedition. He was especially taken with the mount of a bighorn

sheep the men had shot on the upper Missouri River. When he arrived on the lower Columbia the following year, he began asking fur trade workers what they knew about those remarkable sheep, and still had them on his mind when he visited Jaco Finlay at Spokane House in May 1826.

> I made inquiry about a sort of sheep found in this neighbourhood, about the same size as that described by Lewis and Clarke, but instead of wool it has short thick coarse hair of a brownish-grey, from which it gets the name of Mouton Gris of the voyageurs. The horns of the male, of a dirty-white colour, form a volute . . .

Rocky Mountain Sheep
"The horns I purchased for a few trinkets and a little tobacco; they are now in the Museum of the Zoological Society."
—David Douglas

Rocky Mountain Sheep

Rocky Mountain Bighorn Ram

"McKay made us some fine steaks . . . laid on the clean mossy foliage in lieu of a plate, and supping it with spoons made from the horns of a mountain sheep."
—David Douglas

Finlay assured Douglas that the animals lived in the high country to the north, and that he and his sons would try to procure an example for him when they traveled there in the fall. Douglas harbored hopes of shooting one himself, and during his time at Fort Colvile, he searched up the Kettle River, marked on some early maps as Sheep Rivulet, probably with the *mouton gris* in mind. He returned empty-handed, but later that summer he chanced upon the prize he sought at the great trade bazaar of the Dalles. "Here I purchased a pair of horns of a male grey sheep of the voyageurs, for which I paid three balls and powder to fire them," Douglas wrote. "The Indian had the skin dressed, forming a sort of shirt, but refused it to me unless I should give him mine in return, which at present I cannot spare." In a paper on the mountain sheep he delivered to the London Zoological Society two years later, he included voyageur accounts describing the bighorn's range in the Cascades, as well as their value among a tribe that he never personally met: "The Horns are generally converted by the Snake Indians into bows, spoons and cooking utensiles."

Douglas had less luck acquiring a specimen of another desirable quadruped that inhabited the high country. When he visited Fort Colvile in spring of 1827, he found John Work busily preparing the skins of a coyote, a buck and doe mule deer, as well as a matched pair of pelts with pure white fur, short horns, and distinct beards. Douglas referred to these creatures with the voyageur term *mouton blanche* (white sheep), but their color and characteristics make it clear that he was speaking of mountain goats.

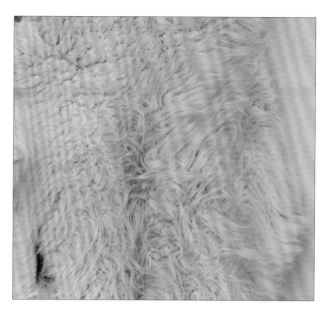

Goat Fleece
"I purchased a little wool of Mouton blanche as a specimen of the quantity of the wool . . . (Get a pair of stockings made of it.)"
—David Douglas

Douglas never did see a live mountain goat or bighorn sheep, but he was certainly keeping a sharp eye out for them as he traveled with the spring express up the Columbia from Kettle Falls into the territory of the Lakes (Sinixt) people. Just north of the 49th parallel, he caught a whiff of an even larger mammal when he saw outsized crescent tracks on the gravel bars of the river. He recognized these first as "reindeer," but quickly picked up the voyageur term "cariboux." Arctic travelers were familiar with the barren ground caribou, but agent Alexander McLeod explained to Douglas that there were two varieties of the animal; the woodland version, which lived on the upper Columbia, was a bit smaller, with "the hair a little darker colour and somewhat curled on the belly and inside of the thighs." Douglas, delighted with this news, deduced what details of life history he could from the tracks he had seen. "The large hoof which this species has (not observed in any other of the genus) is a proof of the wise economy of Nature, given it to facilitate its tedious wanderings in the deep snows."

When he counted "not fewer than a hundred skins" of this reindeer on the floor of a single Lakes lodge, he was observing a dynamic between Sinixt culture and woodland caribou that had been developed over many generations, and that fluctuated over time, season, elevation, climate cycles, and need. The message from those hides would soon be muddled

TAB CCXIV

Carex Douglasii

Douglas's sedge
Carex douglasii
Douglas noted that sedges were used for cordage, basketry, and mats as well as food for both people and animals.

by a host of new variables, and the caribou that in Douglas's time wintered in large numbers around the Columbia's Arrow Lakes while feeding on wetland sedges are today a rare mountain species surviving solely on lichens. People acted as the agents of change in this equation, but in so many nebulous ways that the idea of restoring woodland caribou to their former range cannot even be considered—they live in a different universe now.

That is true of many animals Douglas saw, large and small. He never counted an elk in the Columbia's interior, but twentieth-century introductions have rendered them common over habitats from desert to ridgetop throughout the Intermountain West. Although never encouraged by similar breeding programs, moose today regularly crash though willow riparian stands along the Spokane River, where Douglas spent much of the summer of 1826 without encountering a single one. The mountain beaver that he never saw in its isolated coastal habitats now wreaks havoc in flowerbeds inside the Seattle city limits. The condor that thrived on spawned-out salmon along the lower Columbia has retreated, gone extinct in the wild, risen again as a zoo attraction, and someday soon will be released back on the lower Columbia, thanks in part to data that Douglas collected from local tribal and fur trade sources.

Just as taxonomists of succeeding generations have lumped and split and decoded and changed the names of many of the animals preserved in their museums, so the life history of each of the creatures Douglas described during his Northwest excursions has traveled some mysterious

route to arrive at the niche each occupies today. As Douglas's carefully observed accounts make clear, it is far from easy to predict where the path of any of those species, including our own, might lead next.

Columbia white-tailed deer
"Mr. Douglas brought home the horns of a full-grown male."
—John Richardson

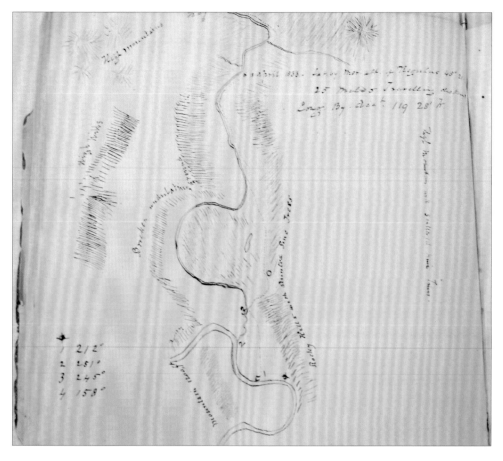

Douglas's Survey Notebook
The first sketch map shows coordinates for Fort Okanagan.
"The Interior, north of the Columbia, is a beautiful and varied country, and
well worth looking after."
—David Douglas

IX.

THE IRON SPHERE

Earth's Magnetic Pulse

∞

BY A FORTUNATE CHANCE OF TIMING, David Douglas's decade of work as a plant collector allowed him to absorb the imprints of some of the most important figures of contemporary exploration and science. In the Northwest, he could trace where he had been, and where he wanted to go, on the maps and routes of George Vancouver, Lewis and Clark, Alexander Mackenzie, and David Thompson. The collector stopped at the Galapagos a decade before Charles Darwin arrived, and climbed Hawaiian volcanoes just after Charles Lyell's *Elements of Geology* had explained their basic properties. As Douglas filled his wooden trunks with specimens and proceeded on to fresh landscapes, he began to think on a larger scale, and described plans to further the scientific expression of his ideas. He had always been interested in everything around him, and now aspired to be a geographer in the broadest sense of the word.

Alexander von Humboldt and Aimee Bonpland at Mount Chimborazo, Equador

A naturalist with broad interests, Von Humboldt studied the relationship between botany, geography, magnetism, and geology.

A scientist of the previous generation who had accomplished just that was the German explorer and polymath Alexander von Humboldt (1769–1859), who had collected in the New World between 1799 and 1804. Humboldt not only carried out major explorations between the Amazon and Orinoco river basins, but also managed to accurately estimate the population of Cuba and fit in an extended stay with Thomas Jefferson at the White House along the way. He returned to Europe with such a wealth of information that it took twenty-one years and thirty volumes to present it all in published form.

Humboldt's approach sounds strikingly modern today. He tried to gather as much unbiased data as possible, then look for organic connections to reflect what he called "the unity of nature." Disciplines that were considered separate, such as biology, meteorology, and geology, were regarded as parts of a larger whole by Humboldt; his aim was to examine them closely enough to reveal the inner workings of the world. His carefully presented results, which influenced scientists from Peru to Siberia, marked the beginning of biogeography. Charles Darwin called Humboldt "the greatest scientific traveler who ever lived," and acknowledged him as a major inspiration for his own seminal theories.

Humboldt always carried a kit of the latest scientific instruments, and took copious astronomical and climatic measurements. In 1817, he introduced his notion of "isothermal lines" (parallel waves that circled the globe), which would allow him to compare temperature and climatic conditions between one land mass and another. Although David Douglas never specifically referenced Humboldt's work, he echoed these ideas when he wrote to a friend about characteristics of plants he had seen during his trip around Cape Horn to the mouth of the Columbia: "it would be interesting to compare the geography of the plants found in the lowest extremities of the Southern continent with the very extensive & beautiful flora of the northern in similar situations."

In 1827, the same year that Douglas returned to London to catalog his Northwest collections, Humboldt retired to the Prussian capital of Berlin with a specific project in mind. While traveling, he had, like Edmund Halley before him, recorded variations in the Earth's magnetic fields from the poles to the equator. In 1804, he presented a paper on the phenomenon that garnered a firestorm of attention—for many physicists of the time, the puzzle of magnetism marked the greatest question left unsolved by Isaac Newton's equations.

Humboldt decided to establish a personal observatory in Berlin and concentrate on investigating these "magnetic storms." He envisioned a network of such observatories, both on land and at sea, that would systematically record the intensity of the magnetic fields, then display their variable patterns as isolines that circled the planet. Through his reputation and personal charisma, Humboldt inspired cooperation among a variety of scientists who agreed with his plan. Interest ran particularly high in England, where the Dollond company, a leading instrument manufacturer in London, began producing high-quality magnetic measuring tools. It was also in England that a former British naval officer named Edward Sabine assumed the mantle of coordinator for the international effort to make use of these instruments.

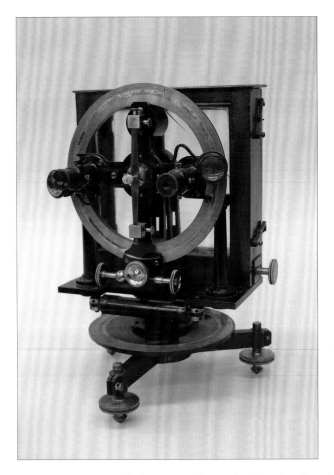

Precision Dip Needle

"Apparatus for Dip and Intensity and variation of the Magnetic needle have been given to me in the most handsome manner from the Colonial Office."
—David Douglas

Sabine had served with the Royal Navy during the War of 1812 and then, after an early retirement, devoted considerable time to the study of physical geography and the properties of terrestrial magnetism. He was elected a fellow of the prestigious Royal Society, which led to an appointment as astronomer on Captain John Ross's first Arctic expedition in 1819. Sabine had chafed at Ross's egotistical style of leadership, and upon their return fell into a bitter dispute with his captain over credit for the magnetic measurements Sabine had faithfully recorded throughout the voyage. While this row caused grumblings among some Royal Society members, Sabine's experience at sea and mathematical background made him a natural fit for Humboldt's master plan.

At this point, David Douglas entered the picture. While sharing lodgings at Chiswick, Douglas found Captain Sabine to be stimulating company, and in 1828 began assisting him in taking geomagnetic measurements on the grounds. The collector must have seen the value of such experiments, because soon after he secured approval for a return trip to the Northwest, he reached out to Sabine for some practical advice. "While preparing for his departure in the summer of 1829, I heard [Douglas] frequently express his regret that his limited education prevented his being able to render those services to the geographical and physical sciences," Sabine later wrote. "He spoke with particular regret of his inability to fix geographical positions."

Recognizing Douglas's combination of seemingly boundless physical energy with "vigor of mind," Sabine offered to serve as Douglas's tutor during the three months that remained before his scheduled departure. At the Royal Observatory in Greenwich, the two fell into a training regime as demanding as any military exercise. Based on his own practical experience, Sabine had a sense of how to feed his pupil "just so much knowledge of plane and spherical trigonometry, and of the nature and use of logarithms, as was essential for his practical purposes . . . For eighteen hours a day he bent all the powers of his mind to overcome difficulties, for which his previous education and habits had so little prepared him."

Edward Sabine
"Capt. Sabine kindly took Douglas under his own particular instructions, so that the latter was soon taught the use of astronomical instruments and became an accurate observer."
—George Barnston

Because of Sabine's deep interest in the earth's magnetic properties and Humboldt's plan for global data collection, Douglas received a very individual mode of astronomical instruction. The captain demonstrated how iron bars suspended from a tripod allowed an observer to measure "the relative intensity of magnetic attraction in different parts of the earth's surface." He acquainted Douglas with the use of the dip circle, a compass supported on gimbals that pivots on a plane to reveal the angle the magnetic field makes with the vertical. This angle of declination had bedeviled ship's navigators since the beginning of sea travel, and Sabine had developed a strict set of rules for taking readings that compensated for iron objects on board.

He also taught Douglas how to hang a magnetically charged needle from a tripod, then watch as Earth's magnetism made that needle vibrate. In London, such a needle suspended from a tripod with special asbestos thread twitched exactly one hundred times in three hundred seconds. By measuring slight differences in the frequency of those vibrations at

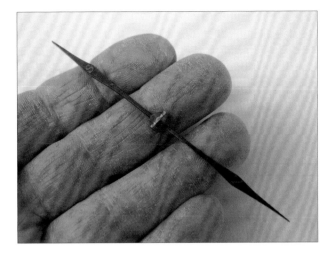

Magnetized compass needle

other locations around the world, Sabine, following Humboldt's theory, had begun to describe sweeping "isographic" curves that covered the earth in exactly the same way "that iron filings arrange themselves around a magnetized iron sphere."

Sabine combined these experiments in magnetism with a thorough knowledge of standard nautical measurements—from the use of accurate temperature and dew point to determine elevation to the intricate trigonometry necessary to calculate latitude and longitude. In addition to the mariner's sextant, Sabine taught Douglas to use a repeating reflecting circle—a larger, heavier instrument that contained two reflecting mirrors and a full circle rather than sixty degrees of arc. The student made good progress throughout this crash course, particularly impressing Sabine with a "capacity which enables him to take in knowledge of various kinds under extreme pressure of time, and to keep each subject as distinct in his mind as if it were the only one of his pursuit." Perhaps Douglas was applying the same kind of self-trained focus that had allowed him, during his time in the Columbia, to separate two puzzling ground squirrels one day and three different lupines the next. Edward Sabine soon felt confident that with a few weeks of steady instruction, combined with the better part of a year to practice on the outbound ship, his pupil would be fully competent "to undertake a variety of determinations which might render his mission important to other branches of science besides those of natural history."

With a Pacific Northwest boundary settlement between British Canada and the United States pending, the Colonial Office in London took an interest in Douglas's newly acquired skills, and supplied him with a stipend for surveying manuals as well as a full set of secondhand but serviceable equipment. Edward Sabine, with a better and more varied set of tools in mind, took Douglas to the shop of Dollond and Sons, the family that had supplied instruments for everyone from Captain Cook and Thomas

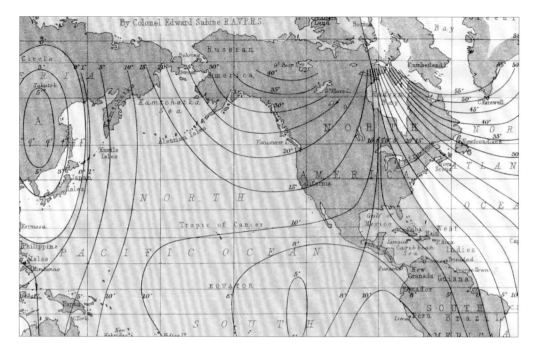

Jefferson to David Thompson. The invoice for the trip totaled close to three hundred pounds, a princely sum at that time. The competent use of such instruments was an ambitious task for a novice, but then his teacher had very ambitious goals in mind.

By 1829, Sabine and Humboldt were well along in assembling their network of field scientists. The latest magnetic theories had been expounded by the Norwegian Christopher Hansteen, and experiments would be carried out by an American expedition under Professor James Renwick and broad-minded sea captains such as the Russian Fyodor Litke. British naval officer Robert Fitzroy, soon to take the *Beagle* around South America with young naturalist Charles Darwin aboard, would also contribute to the project. David Douglas, on his third foray as a horticultural collector, was now assigned to cover a significant portion of North America, and expected to make a very real contribution to the undertaking.

> *Mr. Douglas will be well provided with instruments, and is practiced in the modes of observation. He hopes to determine the magnetic phenomena, from California in the south, to the farthest*

Terrestrial Magnetism
"The accompanying sketch of the northern hemisphere, may enable me to convey a more distinct notion of the arrangement of the isodynamic curves."
—Edward Sabine

extent towards the north, to which circumstances may enable him to prosecute his researches; and from the ocean on the west, occasionally to the Rocky Mountains on the east.

To help his student meet these ends, Sabine provided him with printed forms to fill out. Over the course of the next five years, returning ships dropped off these completed forms for Sabine to use as data for his refined calculations. According to Edward Sabine's final tally for the years 1830–35, he received from Douglas:

3 volumes of science observations

3 volumes of chronometrical observations

3 volumes of magnetic variations and intensity

3 papers of magnetic dip observations

4 volumes of observations of latitudes

1 volume of field sketches

7 papers of meteorological observations

As he progressed around the Horn on his 1829–30 voyage to the Columbia, Douglas diligently practiced with the tools of his new trade. "How beneficial it is for a person like me to be at sea some months previous to engaging on a long journey," he wrote to Sabine. "I have had time to think over and settle my plans; and I never suffered an opportunity to pass without endeavoring to perfect myself in the use of some instruments." He developed a particular fondness for the dependability of Chronometer No. 201, and explained in detail how the reflecting circle lent itself to his present skill level—he found that the extra weight of the circle made it easier to bring his arm and eye together steadily down on the horizon, and believed it helped him take a more accurate shot.

Hudson's Bay Company clerk George Barnston happened to be present at Fort Vancouver when Douglas arrived on June 3, 1830, and he described with great delight the careful unpacking of his friend's new tool kit. Barnston, who had trained as a surveyor and engineer, recognized the quality of the instruments, noting that "even the famous Arago had furnished approved asbestos thread for suspending the magnetic bars." (François Jean Dominique Arago, a French astronomer and physicist known for his experiments in chromatic polarization and electromagnetism, would soon be named director of the French Observatory.) Barnston appreciated Douglas's meticulous recalibration and trial of each device as he demonstrated his mastery over the full range of his new discipline. "His astronomical work advanced surely and rapidly.

Handheld Dip Needle

The regularity of barometrical and magnetical figurings was conspicuous, and the diurnal variations of temperature remarkably equal, the humidity of the atmosphere generally a mere trifle," Barnston wrote. The clerk was equally impressed with his friend's diligence as he calculated shot after shot for longitude, using two different methods of the time, to determine the exact position of Fort Vancouver, which would serve as his benchmark for future observations.

Traveling upstream to Fort Walla Walla, Barnston assisted Douglas in surveys at the confluence of the Snake and Columbia Rivers, a place of great importance to any British claims for the proposed international boundary settlement. As they set up their instruments, the pair would have been well aware that they were repeating observations taken by the Corps of Discovery in 1805 and by David Thompson in 1811. At Fort Walla Walla, Barnston outfitted Douglas with horses and a guide for a summer trip into the Blue Mountains, where the naturalist hoped to begin his correlations of plant ranges with accurate geographical data.

When Douglas traveled back downstream, the portage around the long series of rapids stretching from Celilo Falls through the Dalles allowed him more time for surveying, and he immediately related his accomplishments in a letter to Barnston. Back at Fort Vancouver, chief factor John McLoughlin realized that the size and weight of Douglas's kit required

Entering the Dalles

"You may look upon 121° 07' 07" as a very close approxima-tion to truth, for the longitude of the upper throat of the dalles."
—David Douglas

extra help, and offered the services of a former Royal Navy sailor named William Johnson as an assistant. The two set off on a surveying run through the Willamette Valley, and Douglas apprised Barnston of his progress in a jaunty letter filled with references that only a fellow surveyor could understand, such as the boiling of mercury to remove air bubbles and other impurities, and the instrument-lover's mecca of Dollond and Sons.

> *I have arranged my barometer every way to please me, but I wish you had been with me to have lent me a hand, for I had some trouble boiling the mercury in the tube. Fortunately I can find only .004 of an inch of index error, from the comparison I made with it and my other at Greenwich. I could have done no more, had I been in Dollond's shop.*

Edward Sabine had hoped that Douglas would be able to take magnetic observations along overland fur trade routes through Oregon and into California, but an epidemic of intermittent fever, coupled with an erosion of tribal relationships, convinced the naturalist to travel by sea instead. He repacked his instruments, and with William Johnson still in tow sailed to Monterey in late 1830. Although Douglas's journal for his time in California has not survived, his plant collections and the survey notebooks he sent back to Sabine indicate that he followed a vigorous schedule—shipping almost exactly as many plant specimens back to England as he did during his first tour of the Columbia, and compiling a list of survey coordinates stretching from Santa Barbara to Santa Rosa.

Bristlecone fir
Abies bracteata
"The singular thistle-like cones with the transparent rosin glittering from the long bractea gives a most imposing appearance."
—David Douglas

This list, which displays positions calculated with reasonable accuracy for a person with only two years' training in a challenging discipline, also provides clear snapshots of Douglas in the field. While waiting for a travel visa from the Mexican authorities, he restlessly probed the limits of Monterey Bay, sending back coordinates for the Points of Carmel, Cypress, Piños, and Año Nuevo. He took readings from Cerro de Buenaventura in the hills above Monterey, as well as from several other peaks in the Coast Range. Most of the points coincide with exciting botanical discoveries, such as the Santa Lucia or bristlecone fir (*Abies bracteata*).

When working in the field he usually boarded at Franciscan missions, and easily ignored the disputes that enmeshed many Protestants who visited the Catholic fathers. "I had no difficulty from the beginning with them, for saving one or two exceptions, they all talk Latin fluently," Douglas wrote to Barnston, evoking the language of botanists before attaching one of his characteristic jokes. "Though there be a great difference in the pronunciation between one from Auld Reekie [Scotland] and Madrid, yet it gave us but little trouble. [The Franciscans] know and love the sciences too well to think it curious to see one go so far in quest of grass."

Although his route can be only roughly traced by the dates on plant specimens and the coordinates on his survey sheets, Douglas reached as far north as San Francisco Solano Mission (near modern Santa Rosa) at 38° 29′. He also recorded the coordinates for Fort Ross, the Russian

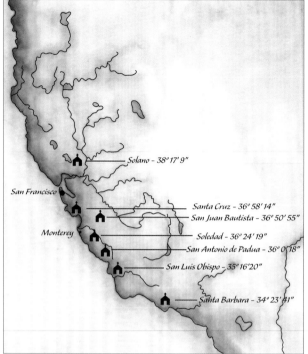

Map labels:

Solano - 38° 17' 9"

San Francisco

Santa Cruz - 36° 58' 14"
San Juan Bautista - 36° 50' 55"

Monterey

Soledad - 36° 24' 19"
San Antonio de Padua - 36° 0' 18"
San Luis Obispo - 35° 16' 20"

Santa Barbara - 34° 23' 41"

Latitudes of California Missions Measured by Douglas 1830-1832

fur trade settlement on the coast northwest of Solano at 38° 51', but he may have learned that position from the head of the Russian-American Fur Company, whom he met with in Monterey. In either case, Douglas's extensive travels and diligent calculations gave him every reason to hope that "perhaps I may at a future time discuss . . . a treatise on the geographical distribution of plants indigenous to North America generally . . . much material in many departments is now in my possession as a ground work."

The Russian captain Fyodor Litke, whom Douglas had met in Greenwich, had encouraged the naturalist's long-held dream of traveling to Asia to compare its terrain with that of North America. "What a glorious prospect!" the excited collector wrote to Hooker. "Thus not only the plants, but a series of observations may be produced, the work of the same individual on both Continents, with the same instruments, under similar circumstances and in corresponding latitudes!" While in California, Douglas received further assurances from the head of the Russian-American Fur Company and from Baron Wrangell, the governor of Alaska, that they would provide passage to the Siberian Peninsula if Douglas could make his way to their headquarters in Sitka.

When he returned to Fort Vancouver from California in the spring of 1833, he had his sights set on a northerly excursion that combined several of his stated goals. Traveling with a fur trade brigade north from Fort Okanagan to Fort Kamloops, then west to the Fraser River, Douglas and William Johnson arrived at Fort Alexandria in early May. This was the place where in 1793 another of his heroes, Sir Alexander Mackenzie, had broken off from that wild river to follow a tribal trail to Bella Coola.

Douglas stayed with the brigade as they continued on by canoe, stopping at the Fort George post (at the modern city of Prince George) only briefly before bending northwest to Fort St. James on Stuart Lake.

While Douglas had written to Hooker that his plan was to "proceed northward, among the mountains, as far as I can do so with safety, and with the prospect of effecting a return," at Fort St. James he decided to go back the way he had come. Accompanied by his assistant, he borrowed a canoe and headed downstream on the Fraser. When they wrecked in a set of challenging rapids south of Fort George, Douglas lost his botanical collections and journal, but managed to save his instrument kit and surveying notebooks. One of the notebooks contained a set of sketch maps depicting his route from Fort Okanagan to the junction of the Fraser and Quesnel Rivers, as well as a series of coordinates attached to locations that extended farther north. The latitudes and longitudes entered new landmarks onto his magnetic globe.

> *Descent from Mountains on Thompson River*
>
> *Bonaparte River, South end of Red Clay Hills*
>
> *1 Mile West of Western extremity of Horse Lake*
>
> *Fort Alexandria*
>
> *Junction Fraser River Quesnel River East Bank*
>
> *Yarechoes village and lake*
>
> *Hudson's Bay House Stuart Lake*
>
> *Fraser River foot of Rocky Islands*

Douglas's longitude reading for Fort Alexandria is particularly accurate, and the foot of the rocky islands in the Fraser River corresponds with Fort George Canyon, also known as the Red Rock Rapids, the scene of his canoe upset.

After refitting at Fort George, Douglas and Johnson began their long trip back to the lower Columbia. The collector detoured to make one more trek into the Blue Mountains to try and recoup some of his lost specimens before continuing on to Fort Vancouver. When an opportunity arose in October, Douglas boarded a ship bound for Hawaii.

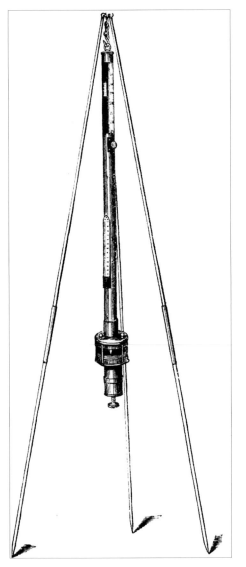

Mountain Barometer

"I have arranged my barometer every way to please me."
—David Douglas

In a letter to Hooker written from the islands, Douglas described himself as much broken down by the ordeal on the Fraser, but added that "I am, thank God reconciled to my loss." Then, as if to prove he was over the mishap, he rattled off five more pages of rapturous prose concerning the wonders of Hawaii. While plants were the focus of any letter to Dr. Hooker, Douglas communicated his enthusiasm for volcanic geology, as well as his mountaineering accomplishments in ascending both Mauna Kea and Mauna Loa, the two highest peaks on the Big Island. Native Hawaiians had told Douglas that one of the peaks had not been conquered since his and Hooker's mutual friend Archibald Menzies scrambled to the top four decades before.

Douglas seemed even more proud of the fact that he had determined the height of both great Hawaiian peaks. "Four feet below the extreme summit of the peak, the barometer was instantly suspended, the cistern being exactly below, and when the mercury had acquired the temperature of the circumambient air, the following register was entered at 11 h. 20 m.; bar 18.362 in.; air 33°; hygr[ometer]. 0.5." By comparing his readings with ones made by an American missionary from his home in Hilo at the same hour, Douglas had acquired the data for a good measurement of altitude. In a letter to Edward Sabine, Douglas organized his barometric readings into sets and interpolated the results according to two accepted methods of calculation. His results came very close to the accepted elevations of Mauna Loa and Mauna Kea today.

But Douglas's letters always contained more than raw numbers. During the arduous climb up Mauna Kea, he observed that large acacias, used by local people for their dugout canoes, appeared only at a band fifteen hundred feet above sea level. He noted that, unlike the tall mountains around

Athabasca Pass in the American Rockies, ambient sound on Mauna Kea was "but very slightly diminished at the summit, owing, undoubtedly, to the absence of snow." When determining the dew point via a touchy process of applying ether to a standard ball, "the immense difference at the summit instantly caught my attention: and fearful lest the ether might have been applied too copiously to the ball, I repeated the experiment five times, with always the same result. So delicate was the dew-ring seen, that it appeared like an exceedingly fine grey silk thread: yet the moment of its appearance was readily perceived." When Douglas calculated latitude both by taking observations with a sextant and through the use of his favored reflecting circle, his results agreed within five seconds. Edward Sabine employed expert mathematicians from Greenwich to interpret Douglas's data and rework these coordinates, and according to later papers was well pleased with the results.

Douglas arrived in Hawaii during a time of restless volcanic activity around Kilauea crater, and his trip to the cauldron linked his boyhood interests in plants, birds, and human history with his new passions of surveying and geology. An elder "Priest of Peli (the Goddess of the Volcano)" described for Douglas an eruptive event in the crater that had taken place in 1787 and killed more than five thousand islanders. He also learned that in June 1832, on the neck between two inactive vents, the ground had opened up and discharged molten lava into both craters over a period of three days. The fact that this neck proved to be the same spot where the poet Lord Byron had set up a temporary house when he visited the crater in 1825 only added to Douglas's delight.

Relying on the priest of Peli and his guide, John Honorii, as informants, Douglas recorded Hawaiian words for place names such as Punahala, "broken in," to designate local landmarks formed by previous volcanic action. The letter to Sabine describing this trip is decorated with wiggly parallel lines to represent streams of lava—some pressed forward transversely, others fluted longitudinally. The volcanic processes at Kilauea inspired some of Douglas's most passionate prose, much of it written from a ledge that offered a panoramic view.

> *A space of five miles square, recently in a state of igneous fusion, in the process of cooling has been broken up into immense ledges and rolled masses, like the breaking up of a great river of ice; and these*

Walls of Crater
"At some places the hardened lava resembles the Gothic arches of an immense
building, piled in fearful magnificence."
—David Douglas

*are of every shape and form, from gigantic rolls, like enormous
cables, to the finest threads, like human hair, which are carried by
the wind for the distance of miles around this terrific laboratory.*

Through it all, he continued to take magnetic observations and sextant
shots by day and by night—"The latitude of this tent, by one meridian altitude of the sun, two passages of Sirius, and one of Canopus, is 19° 25' 42"
N." Writing to George Barnston, he tried to explain the effect of his time
on the volcano: "Were the traveler permitted to express the emotions he
feels when placed on such an astonishing part of the earth's surface, cold
indeed must his heart be, to the great operations of nature, and still colder
towards nature's God, if he could behold them without deep humility and
reverential awe. Man feels himself as nothing, as if standing on the verge
of another world."

To Douglas, it seemed as if the mechanisms that stirred his emotions
also moved the physical world, forming grand connections between magnetism and geology. He wrote to Sabine of the regular twitching and jerk-
ing motions in his needle when set up at the crater. "The cause was not
permanent, but very variable; and did not arise from the accidental presence of any mineral substance, but from a sympathy between the *magnetical*
action and that going on in the center of the volcano." Here Douglas was
feeling the subtle quakes that ripple deep within the earth under active
fracture zones, and interpolating them into earth-changing events. "I cannot attempt to describe the sensations felt," he told Barnston, "the even
fearful excitement experienced during my visit to this place."

Such exciting sensations always stirred Douglas's interest. In his last
report to Edward Sabine, in May 1834, Douglas wrote that although he
intended to return to England at the first opportunity, he was far from
being finished with his work. "I shall continue to labour at the islands," he
promised, "to the best of my ability."

Sandstone and Basaltic Rocks
*"From the oppressive heat I found great relief by bathing
both morning and evening."*
—David Douglas

X.
TRAVELERS

Riding the Wind

❦

THE FIRST WEEK OF AUGUST, 1826, found David Douglas plodding beside a hundred freshly traded Nez Perce and Palus horses, as the Hudson's Bay Company drovers he was accompanying made their way from the Snake River north to the Spokane country. Two familiar friends, John Work and Finan McDonald, led the drive along a tribal road that Douglas found tolerable in some stretches, but "in others very bad, with badger and rabbit holes, which at this season are covered with grass, rendering them more dangerous."

Aspen Camp

"The whole country is destitute of timber; light, dry gravelly soil, with a scant sward of grass."
—David Douglas

Over the course of two weeks spent among the rugged basalt flows of the lower Snake, the naturalist had focused considerable effort on plucking and identifying some of those same native bunchgrasses, and within the dust cloud stirred up by the horse herd he continued to circle the brigade like an anxious puppy, searching for the ripe seeds of flowers that had caught his fancy earlier in the season. In a region "destitute of timber," under the additional burden of hundred-degree heat, he picked up a couple of new species of milk vetch and two more monkeyflowers. But even Douglas's enthusiasm had its limits, and when the party camped at a stagnant pothole beneath a lone cottonwood tree, his only comment about the yellow water lilies and green pondweed splattered across the pond's surface was that they made the "water very bad."

Douglas's spirits lifted as the brigade approached the Spokane River under shady belts of ponderosa pine. After a side trip to the refreshing cascade of Spokane Falls, he made his way back downstream, harvesting seeds and bulbs of three of his favorite lilies, known today as yellow bells (*Fritillaria pudica*), glacier lily (*Erythronium grandiflorum*), and sagebrush mariposa lily (*Calochortus macrocarpus*). Despite the stunning difference in soil and climate, he had high hopes for their prospects as garden plants in Great Britain.

———

Summer winds today blow at least as hot on the eastern fringe of the Columbia Basin as they did in Douglas's time. Wheat and lentils cover the rounded hills where wind-blown glacial soils have settled, but travel routes have always worked their way through the network of channeled scablands scoured by Ice Age floods. Although David Douglas, even in the harsh conditions of August, recorded a wide variety of grasses along with a few persistent wildflowers, it would be interesting to know how much the coming of horses, less than a hundred years before his own arrival, might have altered the mix of vegetation that he saw.

In the two centuries since he passed through, the Columbia Basin has been assailed with familiar names from the noxious weed list of the American West: cheatgrass, Russian thistle, tumblemustard, knapweed, and many more. In certain areas, with a certain frame of mind, it feels like there is nothing else to see. Yet, along the trails that Douglas followed, glimpses of the world before wheat and weeds are still in evidence.

Just east of the shadow community of Ralston, Cow Creek sleeps beneath an ocean of tule rushes stout enough for the tallest tepee mats. On either side of the watercourse, wide bands of reed canarygrass, far too dense to be a native plant, follow the creek's windings. Beyond those noxious bands, however, the canarygrass gives way to alkaline flats dimpled with the rocky hummocks that make *scabland* a pleasant word. Bright green greasewood and the red-tinged scales of hopsage announce their presence, and stiff sagebrush, colored an intense olive green and dripping with scent, crawls across the rocks at ankle height. Smaller clumps of buckwheat and phlox, perfectly adapted to ten inches of rain a year, have anchored themselves to fractured basalt. White patches of volcanic ash, left over from Mount St. Helens's 1980 eruption, surround the dark lithosols, supporting strong tufts of Idaho fescue and bluebunch wheatgrass, junegrass, Indian ricegrass, and wispy needle-and-thread. The alkaline swales burst with big sprays of Great Basin rye. This is the same list of native bunchgrasses that Douglas collected here, and the open space that comfortably separates the plants comes crusted with desiccated lichens

Sagebrush mariposa lily *Calochortus macrocarpus*
"I observed it in abundance, on the banks of the southern branches of the Columbia."
—David Douglas

and mosses, the official anchors of healthy shrub-steppe habitat.

I wind through this vast rock garden for a couple of miles before spotting a slender naked stem that sprouts from a circle of ash. It's a bloomed-out sagebrush mariposa lily (*Calochortus macrocarpus*), capped with a seed pod the color of baked kaolin clay. The pod consists of three dried sepals closed like a tulip, full of irregular peppercorn seeds that glow with the obsidian and sorrel colors of the desert. When a gust of wind flaps up, the seeds left inside the pod rattle like bones.

That small racket is echoed through the bunchgrass by dozens, then hundreds, of skeletal lily stems, each one rattling its song. Now I understand how Palus, Coeur d'Alene, Spokane, and Columbia Salish bands could have shared in such a wealth of good food. Douglas carefully packed a selection of those seeds in folded paper, adding them to a tin box with hundreds of other collections that he would carry back to England. He described his prize to the Horticultural Society in 1828, only a few months after his return: "The species of *Calochortus* are bulbous rooted plants, with striated stems, narrow sheathing leaves, and beautiful purple, or white flowers, remarkably bearded in the inside." As he read his paper, he could admire the mature blooms of sagebrush mariposa lilies propagated in Chiswick from the seeds he had collected alongside that dusty horse brigade. He could also

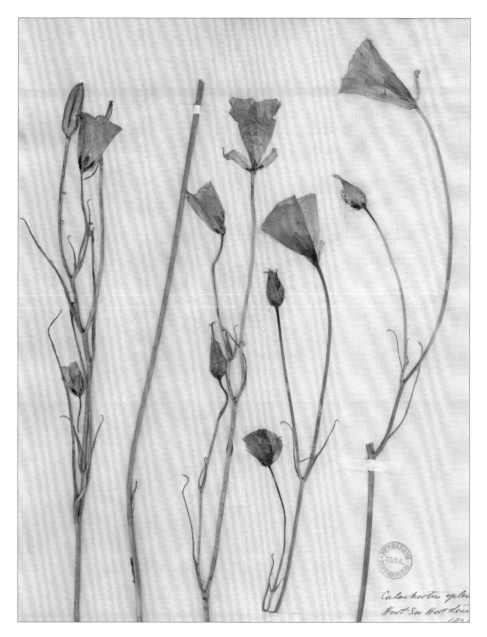

Specimen Paper of Splendid mariposa lily
Calochortus splendens
"I have doubled the genus of Calochortus."
—David Douglas

hope that two additional mariposa species he had brought back would soon flower beside those purple sagebrush lilies.

When Douglas returned to North America a year later, he moved south to California and doubled the number of known *Calochortus* species, sending more viable seeds back to England. The Chiswick nurserymen knew exactly what to do with this new trove of closely related stock, and succeeding horticulturalists have followed suit. Growers and seed catalogs have created a whole industry from the *Calochortus* genus, developing hybrids to emphasize desired traits. Still, many purists prefer the native species, whose common names reveal their timeless attraction: green-banded star tulip; elegant mariposa lily; fairy lantern; white globe lily; naked cat's ear.

———

Even before Douglas touched land in the Pacific Northwest, Europe had been launching some of its own plant wealth across to the New World. Alien seed borne in livestock feed and company gardens had already begun to creep along the river bottoms, and Douglas recorded several introduced plants among his early collections. Each species constituted one unpredictable new element that over time would establish its own relationship within the unfamiliar surroundings. While Chiswick nurserymen took great care to manage the boundaries of new mariposa lily cultivars in their garden, these introductions to the New World had an entire hemisphere to test for compatibility.

Chenopodium is a genus of plants native to Eurasia whose roughly triangular leaves resemble goose feet—the exact meaning of their Latin name. *Chenopodium album* carries tiny scales that give its leaves and stems a dappled grayish-white cast. *C. album* is a very adaptable plant; its form can vary from a thick-stemmed six-foot soldier to a loose open lattice. The tiny grayish-green flowers bloom in great clusters from the top, reminding many people of purple amaranth. These blooms produce tens of thousands of nutritious groat-sized seeds.

Ancient Romans, probably referring to the late-summer ripeness of these seeds, called the plant *lammas quarter* after their August harvest moon, which in turn gave rise to the common name lambsquarters. English country folk have long called it fat-hen or mealweed because its odorless leaves and seeds are good for fattening up chickens. Their kin carried it to America, where hungry hogs ingested whole plants and then spread seeds across the continent. I learned of it as pigweed and hated to hoe it out of gardens, first in the Carolinas and then in northeastern Washington—a weed just as vigorous in the 90 percent humidity of the South as it is on bone-dry August days in the Inland Northwest.

Any plant that has managed to travel from the Mediterranean across Africa, Australia, the most remote Pacific Islands, and all of North America must have a close connection with humans, and both the leaves and seeds of lambsquarters have been widely eaten by people as well as animals. *Chenopodium album* has been purposefully cultivated in India for millennia, and in Great Britain, several Neolithic archaeological digs have found lambsquarters seeds preserved in earthenware pots. Tollund Man, an Iron Age body preserved in a Dutch peat bog, had it as cereal for his last breakfast. Over many centuries, the tradition of eating these seeds held strong. Field cooks for Napoleon Bonaparte (who always paid close attention to the diet of his troops) used ground *Chenopodium* seeds to supplement wheat and barley flour for his army's endless loaves of bread.

During his childhood, David Douglas probably dined on lambsquarters greens, and when he ran across the plant on the lower Columbia, he recognized it immediately. Collection number 410 in his summer 1825 field journal noted: "Chenopodium sp.; annual; plentiful in all rich soils; fertile banks of rivers; grows very strong and abundant around Indian villages and camps." What confounded the collector, however, was that Chinook people, whom he knew to have a wide range of botanical knowledge, should ignore such an obvious food source.

> *Among the numerous vegetables used by them, it is rather singular they should omit one which is almost universally used in every country; even the tender shoots of several species of Rubus* [currants, salmonberries, and thimbleberries]*, and sprouts of a species of Equisetum* [horsetails] *and Scirpus* [rushes] *are greedily sought after, used and considered good, while this wholesome plant is left untouched.*

Of course people tend to eat what they are accustomed to, and *Chenopodium* would have been a very recent arrival in the Chinookan world. But people are adaptable, and as the plant spread across the

Lambsquarters or Pigweed
Chenopodium album
"*Near Indian villages; common.*"
—David Douglas

Northwest, at least some tribes learned to like it. Almost exactly a century later, an anthropologist named James Teit again recorded the presence of lambsquarters. Teit married an Interior Salish woman (of the *nlaka'pmx* or Thompson tribe, based in the Thompson River country of Interior British Columbia), and had extensive experience with tribal plant usage. As a man of British upbringing, he referred to *Chenopodium album* as lambsquarters instead of our more vulgar pigweed, but they are both the same troublesome or nutritious plant, depending on how you look at it. "Common lambsquarters was called 'stuwituimax or stuwituimax a sama' [sama is a Salish word for white men]," wrote Teit. "[They are] weeds introduced by whites, a plant with no particular name or use especially annuals . . . [lambsquarters] now used as a green boiled no doubt learned from whites . . ."

Teit made no mention of any tribal use of *Chenopodium* seeds, despite their long history as an important source of nutrition on several continents. As Douglas amply documented, there are plenty of sunflower and biscuitroot family plants in the north Columbia country that produce bigger, tastier seeds, and these have remained a preferred food for many tribal families. After two centuries, some might view lambsquarters as a newcomer to their world; others, who travel the globe alongside the plant, might think of it as a familiar old friend.

The thick stands of reed canary-grass (*Phalaris arundinacea*) that crowd the course of Cow Creek around Ralston look all too familiar to a modern eye. Locals compare it to bamboo because of its light green color and long wide leaves that spread at right angles from swollen stem joints. In late spring, its new flowers give the plant a purple tinge that fades to light straw by summertime. This head-high grass sprouts an even taller seedhead that rains a shower of fine down and pollen dust on anyone foolish enough to walk through a stand in the summer months.

Cock of the Plains (Sage Grouse)
"The pheasant-tailed grouse nests on the ground, beneath the shade of Purshia [bitterbrush] or Atemesia [sage], or near streams, among Phalaris arundincea [reed canarygrass]."
—David Douglas

Although reed canarygrass has a long history in the annals of botany, it remains little understood. It is usually regarded as a native of northern Europe; varieties were cultivated in England as early as 1824, and in Germany soon after that. By the 1920s, European seed companies were exporting seeds of cultivated *Phalaris* varieties to America.

All of this sounds like the common story of a benign plant whose introduction to a new continent has resulted in an invasive weed. But the twist here is that David Douglas collected specimens of *Phalaris arundinacea* from the Columbia River in what appear to be completely native settings. Even more surprising is his mention of the grass in his account of the sage grouse, which he described as

> *plentiful throughout the barren arid plains of the river Columbia . . . in the summer and autumn months these birds are seen in small troops, and in winter and spring in flocks of several hundreds . . . Nest on the ground, under the shade of Purshia* [antelope bitterbrush] *and Artemisia* [sagebrush], *or near streams among Phalaris arundinacea* [reed canarygrass], *carelessly constructed of dry grass and slender twigs.*

Reed canarygrass
Arundinacea phalaris

"This does not seem to differ in any respect from the European plant."
—William J. Hooker

Thus Douglas saw the totemic sage grouse, a bird that has almost entirely disappeared from the Columbia Basin, construct its nest among reed canarygrass, a plant that today has overrun large tracts of wetland habitat all over the Basin and beyond.

Douglas was the first to record *Phalaris* in the Northwest, but his sample was backed up by other collectors in Oregon in the 1840s, and International Boundary Survey naturalists in the late 1850s collected it along the 49th parallel from coastal British Columbia to the Idaho Panhandle. Most botanists today believe that the canarygrass noted by their nineteenth-century cohorts had lived since the ice age floods as one constituent of diverse wetland communities, sharing space with reeds, sedges, rushes, and a suite of other native grasses.

Homesteaders across the Northwest saw the plant as an asset, and worked to grow more. In the late 1800s and early 1900s, many planted reed canarygrass as a "breaking in" crop, broadcasting it around recently logged stumps and debris piles that would be slowly converted to farmland. The grass was not the greatest of forage when green, and did not suit horses at all. But it was productive, could be used for both pasture and hay, and did not seem to be bothered by periodic drought, flood, cold weather, or overgrazing. It was also quick to establish itself in disturbed situations such as burned ground or raw ditchwork.

As more settlers pushed back the woods and drained the marshes of the Inland Northwest, something happened to disrupt the balance between reed canarygrass and other wetland plants. Possibly due to a blending of native stock with more aggressive American or European cultivars, canarygrass began to dominate the native communities. When it became so widespread that many farmers, restoration experts, and wildlife managers wanted to get rid of it, reed canarygrass, like so many well-known weeds, proved to be almost unbelievably resilient.

Equilibrium is a nebulous word in the natural world, and the plant communities of the Northwest were already undergoing dynamic changes when Douglas arrived. He began his compilations exactly three decades after Captain Bishop and the crew of the *Ruby* planted a garden of corn, radishes, and more inside Baker Bay in 1795; about twenty years after the Corps of Discovery wintered at Fort Clatsop and David Thompson established inland fur trade posts upstream. The fact that the collector watched workers lay out new farm fields at Fort Vancouver and shared canoe space with three piglets and three calves on his first trip to the Interior in March 1826 should have provided a clue that lambsquarters, and more plants like it, might soon follow in their wake.

Introduced weeds seem to have the ability to draw wildly different responses from the people who live with them. While some residents might wince each time they pass a knapweed growing by the trail, others can step past hundreds of that species in order to pull a few of the Dalmatian toadflax that are driving them to distraction. Each species loops through its own tapestry of place and time. Some, like reed canarygrass, fade away at some point only to reemerge with demonic force in another. The only common element seems to be that each plant labeled as a weed has grown beyond the control of the humans who introduced it in the first place. Given the number of plants that Douglas carried back to England, it only makes sense that some of his collections might have taken similar advantage of their new environment.

After his mentor Archibald Menzies recommended salal as a fine prospective ground cover, Douglas made it one of his first collections on the Columbia. Salal did prove to be popular among British gardeners, but it was also hardy enough that it soon escaped and began spreading into mild, wet habitats around the islands. Today salal hinders walkers' footsteps from Wales to Scotland, just as it did Douglas's early treks across Cape Disappointment. And while he enjoyed the jet-black salal berries, the fruit never caught on as a popular food in Great Britain—salal remains all weed to most people there.

Salmonberries were another valued Northwest coast food that London palates declared "of no great edibility," but the shrub itself sold well because

Salmonberry
Rubus spectabilis

"The young shoots are stripped of their bark and eaten in a raw state by the natives."
—David Douglas

of its attractive dark pink flowers and strong growth habits. The same vibrant shoots that Douglas watched Chinook people collect in the spring promised quick growth, prompting some estate managers to set out bushes as cover for game. Those young suckers spread to form dense thickets that forced out other native herbs and shrubs, then began displacing young trees and altering faunal regimes until it was too late to stem their tide. Salmonberry now appears on most of Great Britain's "plants of concern" lists.

Then there is *Epilobium angustifolium*, called fireweed in North America and rosebay willow herb in Europe. Native to both continents, it is described as "strongly colonial" in American plant manuals. When western wildfires give rise to entire hillsides of fireweed's vigorous purple blooms, no one seems to mind. In Europe, however, the plant fit a quieter profile until the past century or so, when it began to change its ways. Some botanists remarked on the strangeness of the incident when rosebay willow herb completely covered an area of woodland destroyed by fire on the Isle of Wight in 1909. In succeeding years it sometimes grew in great profusion after burns and clearcuts on closely managed forest land, and some stands, instead of disappearing when a canopy closed, outcompeted other emerging plants. During the early years of World War II, rosebay willow herb exploded out of bomb craters left behind by Hitler's Blitz. No longer confined to short bursts of activity, one closely monitored stand on a Dutch sand dune continued to increase for thirty-five consecutive years.

Agronomists began to suspect that some kind of hybrid vigor must be involved. Lab work revealed that the genetic structure of fireweed introduced from North America contained seventy-two chromosomes, while the native species had been getting by with thirty-six. Such theories exactly mirror North American ideas about reed canarygrass, and they share a common root: both plants were collected in the Columbia River drainage and sent back to London by David Douglas, and both have since changed the way they grow.

———

For some observers, no suite of plants could be more in balance with its surroundings than a stand of native wildflowers and bunchgrasses strewn across an expanse of rocky shrub-steppe. For others, such a scene presents an opportunity to transport some of the alluring blooms to a more manageable garden bed, where they can be tended, crossed, and shared with fellow enthusiasts. It is hard to tell which side of this equation Coeur d'Alene elder Felix Aripa is straddling as we stroll along the edge of a wetland on the Washington-Idaho border, not far from the place where David Douglas collected his own stock of sagebrush mariposa lily seed. Aripa grunts and smiles as he points across standing water to the clean white ends of aspen limbs, freshly gnawed and expertly woven into a large beaver dam.

"My dad always told us to watch the beaver," he says. "See how he goes through the day and night. See how he keeps up with his work. Maybe you'll learn something.

"And now look. They let the beaver return into this marshland, after trapping them out for all those years since the fur traders came. Beaver builds one dam that slows the water down, and things begin to change."

Fireweed or Rosebay willowherb *Epilobium angustifolium* "From Lake Huron and Newfoundland in the east to the shores of the Pacific, on the Columbia River." —William J. Hooker

Plateau Wetland with Reed Canarygrass

This wetland had been dying over the past century, drying up as the creek that ran through it was channelized to create farmland, then drying up more as reed canarygrass encroached on every part of the system. During all that time, any beavers that arrived to work on the cottonwood stands along the creek margins had been systematically trapped. Then, during the course of a recent restoration project, those immigrants were left alone. The beavers constructed a new dam that slowed the channelized creek, just as Aripa described, then spread its water over broad fields of canarygrass. Instead of being seasonally wet, the fields now bathe in standing water year-round. Within a few years of the beavers' return, the canarygrass was in obvious retreat. I think once more of Napoleon's armies, always advancing in waves. There came a time when even they fell back.

As we try to calculate the invisible changes of water flow around us, Aripa lifts one hand into the air and pinches a floating feather between his thumb and forefinger. It is a cattail seed, perhaps released from a patch that lies upwind, on the far side of the burgeoning marsh. He looks at it and smiles, then heels out a space among the tangle of reed canarygrass at our feet and bends down to touch the seed to black mud. He stands up slowly, still talking about beaver chew, and pats the mud spot gently with his heel.

"I used to see the elders do this when I was a kid," he says. "Just helping things along.' That's what they'd say. 'Just giving one traveling seed a chance.'"

We are all travelers, of course. Beavers wander into wetlands, gnaw through stands of cottonwood and aspen that provide both their favored food and raw material for their dams, then move along. Sagebrush mariposa lilies had carved out the largest range of any member of the *Calochortus* genus, flowering from British Columbia to California, long before David Douglas carried their seeds to a new continent. The homestead family that channelized this small creek a hundred and twenty years ago had traveled here from Europe, where such controls have been part of the landscape since medieval times. As that straightened watercourse changed the landscape around them, Felix Aripa's parents slowed their journey through an ancient annual round in order to tend their own newly productive fields on the recently established Coeur d'Alene Indian Reservation.

If Aripa and his Coeur d'Alene ancestors represent many generations of focus on the distribution and effects of planting the seed of a single species, then David Douglas offers a snapshot in time for many hundreds of such travelers, arrayed within the complex of their fellow voyagers. From the moment he touches each plant, we can see a species wax and wane in relationship to the others around it.

Douglas did the work that was asked of him as well as he possibly could. The hundreds of names he bestowed on the flora and fauna in our part of the world in no way changed the plants and animals that still carry them, but they do provide a context as that work is passed on to this generation, then played forward to the next. Whoever happens to be here, the seeds will continue to float past, riding on the wind.

Menzies's larkspur
Delphinium menziesii

CHRONOLOGY

New World Explorations, Science, and David Douglas

1793 Alexander Mackenzie travels overland across the Rockies to the Pacific Coast

1794 Archibald Menzies collects plants with George Vancouver's Pacific Coast survey

1799 **David Douglas is born in Scone village, Perthshire, Scotland**

Alexander von Humboldt gathers biological and geomagnetic data in Latin America

1801 André Michaux's *The Oaks of North America* is published in Paris

1804 London Horticultural Society is established

1805 Lewis and Clark winter at the mouth of the Columbia River

1810 **Douglas begins summer work in the gardens at Scone Palace**

1811 David Thompson completes first survey of the entire Columbia River

1814 Frederick Pursh publishes *Flora Americae Septentrionalis* in London

1818 John Ross and Edward Sabine gather geomagnetic data in the Antarctic

1819 John Franklin begins his first Arctic expedition

1820 **Douglas works under William Jackson Hooker at Glasgow Botanic Garden**

1823 **London Horticultural Society sends Douglas to collect in mid-Atlantic states**

1824 **Douglas departs for the Pacific Northwest**

Franklin publishes *Journey to the Polar Sea*

1825 **Douglas collects on the lower Columbia**

Franklin departs on his second Arctic expedition

1826 **Douglas collects on the Columbia Plateau**

1827 **Douglas travels with a fur brigade across Canada, then returns to England via Hudson Bay**

1828 **Douglas works in London and delivers several scientific papers**

1829 **Douglas studies surveying and geomagnetic measurements with Edward Sabine in London, then departs for second trip to the Pacific Northwest**

William Jackson Hooker publishes volume 1 of *Flora Boreali-Americana*

John Richardson publishes *Fauna Boreali-Americana*

1830 **Douglas stops in Hawaii on the way to the Columbia River**

Charles Lyell publishes volume 1 of *Principles of Geology*

1831 **Douglas travels in California**

Charles Darwin departs with Captain Robert Fitzroy aboard the *Beagle*

1832 **Douglas returns to the Columbia**

Aylmer Lambert publishes the third volume of *A Description of the Genus Pinus*

1833 **Douglas travels north to Fort St. James via the Fraser River, then departs from the Columbia for Hawaii**

1834 **Douglas dies in a cattle pit trap in Hawaii**

Douglas's "Volcanoes in the Sandwich Islands" appears in *Geographical Society*

1836 Hooker publishes "A Brief Memoir of the Life of David Douglas" in the *Companion to the Botanical Magazine*

1837 Edward Sabine presents "Observations taken on the west coast of North America by the late Mr. David Douglas"

1914 The Royal Horticultural Society publishes a selection of Douglas's writings

ACKNOWLEDGMENTS

Grateful thanks to all the people who helped me with this project.

There are far too many to name them all, but the following individuals and institutions provided either specific or long-term inspiration.

Kathy Ahlenslager, Nancy Anderson, Felix Aripa, Joe Arnett, Steve Arno, Louie Bair, Katie Barber, Jean Barman, Dale Beeks, Ted Bennima, Shawn Brigman, Raymond Brinkman, Jennifer Brown, Angela Buck, Rex Buck Jr., Rex Buck III, Pam Camp, Tucker Childs, Jackie Cook, Chalk Courchane, Earl Davis, Denny DeMeyer, Peter Dunwiddie, Thron & Betsey Ellerbroek, Ann Ferguson, Charles Ferree, Geri Flett, Pauline Flett, Vi Frizzell. Bill Garvin, Darlene Garcia, Dr. Meredith Heick, Bill & Diana Hottell. Syd House, Larry Hufford, Gene & Nancy Hunn, John Jackson, Marilyn James, Tony Johnson, Andy Korsos, Lois Leonard, Chris Loggers, Judith Lowery, Gary Luke and the staff at Sasquatch Books, Carol Mack, David Malaher, Dottie Marchand, Melinda Morningowl, Pat Moses, Michael Powell, Klaus Puettmann, Robin O'Quinn, Oliver Rackham, Marsha Rooney, Bruce Rigsby, Tim Ryan, John Schenk, Donna Sinclair, John Stuart, Jane Swiatek, Cindy Talbott, Nancy Turner,

Sylvia von Kirk, Bruce Watson, John Woods, Marian Wynecoop, Marsha Wynecoop, and Henry Zenk

Center for Columbia River History, Columbia Pacific Heritage Museum, Columbia River Bar Pilots, Columbia River Maritime Museum, Conner Museum at Washington State University, Kew Herbarium at Royal Botanic Gardens, Marion Ownbey Herbarium at Washington State University, Native Plant Societies of Idaho, Montana, Oregon, and Washington, New York Historical Society, Northwest Museum of Arts and Culture, Royal Horticultural Society, Slater Museum of Natural History at University of Puget Sound, and Washington State Historical Society

CHAPTER NOTES

Abbreviations:

CBM: *Companion to the Botanical Magazine 2* (1836)

HBCA: Hudson's Bay Company Archives, Winnipeg, Manitoba

PROLOGUE

"A tall splendid tree": Douglas, *Journal* (1914), 143.

"Make a point of obtaining it": Ibid.

I

William and Ann's voyage to the Columbia: Log of the *William and Ann*, 1825, HBCA C.1/1066.

"furious hurricanes of North-West America": Douglas, *Journal* (1914), 101.

"with the anticipation that our voyage would speedily be at an end": Scouler, "Journal," 52.

"Boisterous and frightful weather": Douglas, *Journal* (1914), 101.

"The breadth of the river at its mouth is about five miles": Ibid., 103.

"joy and expectation sat on every countenance": Douglas, "Sketch," *CBM*, 88.

"passed over the sand bank in safety": Douglas, *Journal* (1914), 101.

"I arrived on the 7th of April": Douglas to Booth, October 2, 1825, Hooker Papers.

William and Ann's outgoing cargo: Log of the *William and Ann*, HBCA C.1/1066.

"ran out against a very heavy Sea": Ibid.

"Did you hear of the total wreck": Douglas to Hooker, September 14, 1829, Hooker Papers.

"The ship which sailed with us was totally wrecked": Douglas to Hooker, October 11, 1830, *CBM*, 148.

"I am no coward either in the water or on the water": Douglas, "Sketch," *CBM*, 112.

II

"The sight of land": Douglas, *Journal* (1914), 101.

"the pleasure of again resuming my wonted employment": Ibid.

"with much curiosity and interest": Douglas, "Sketch," *CBM*, 88.

"The natives viewed us with curiosity and put to us many questions": Douglas, *Journal* (1914), 102.

"These characters approximate them": Scouler, "Journal," 163.

"the reception we experienced rendered it prudent to leave": Ibid., 165.

Chinook women and children gathering horsetails: Ibid., 168.

Tribal canoe paddlers snacking on salmonberry shoots: Douglas, *Journal* (1914), 106.

Scouler's description of Comcomly and his village: Scouler, "Journal," 167–68.

"the natives are in general very friendly": Douglas to Booth, October 2, 1825, Hooker Papers.

"I cannot but say he afforded me the most comfortable meal": Douglas, *Journal* (1914), 137.

"baskets, hats made after their own fashion": Ibid., 138.

"my shot carried away all the crown": Douglas, "Sketch," *CBM*, 98.

purchase of "several articles of wearing apparel": Ibid., 91.

"faithful to his proposition": Douglas, *Journal* (1914), 144.

"I think them a good specimen of the ingenuity of the natives": Ibid.

"This old man sent his canoe and twelve Indians to ferry us": Douglas, "Sketch," (1914), 61.

"imitated English manners with considerable nicety": Douglas, "Sketch," *CBM*, 96.

"every kindness and all the hospitality Indian courtesy could suggest": Douglas, *Journal* (1914), 149.

time "during which I experienced more fatigue and misery": Ibid., 150.

"There is in the root a large quantity of farinaceous substance": Ibid., 149.

"My time laying heavy on my hands, I resolved on visiting the ocean": Ibid., 239.

"the house of my old Indian friend Cockqua": Ibid.

"Not twelve grown up persons live": Douglas to Hooker, October 23, 1832, *CBM*, 155.

"Villages . . . are totally gone": Douglas to Hooker, October 11, 1830, *CBM*, 147.

roots of bracken fern were "dried and eaten": Douglas, *Journal* (1914), 137.

Salmonberry: "fruit large, oblong, yellowish-white, and well flavoured": Hooker, *Flora*, I:178.

Salal: "bears abundantly, fruit good": Douglas, *Journal* (1914), 104.

"they are much prejudiced in favour of their own way of living": Ibid., 127.

III

"always a little before day": Douglas, *Journal* (1914), 160.

"the snow lies 3 to 5 feet deep": Ibid.

"Umbellifereae, perennial; flowers purple": Ibid., 168.

"This highly ornamental plant I must try to preserve": Ibid., 163.

Lakes (Sinixt) tribal member Marilyn James: Discussions with the author, January 2011.

requested that his hosts "should dry me the seeds": Douglas, *Journal* (1914), 172.

ways that continue to confuse modern researchers: Hunn, 99–109.

"I get it generally stewed down in a little dried buffalo-meat or game": Douglas, *Journal* (1914), 177.

"in the spring the sagebrush mariposa forms an article of food": Douglas, "An Account of the Species of *Calochortus*," 278.

"Its small roots are eaten by the natives": Douglas, *Journal* (1914), 167.

"The roots of this are gathered in great quantities by the Indians": Hooker, *Flora*, I:223.

The sentiments of Spokane elder Pauline Flett: Discussions with the author, March 2007.

"called by them Missouii": Douglas, *Journal* (1914), 176.

"a sweet taste, resembling that of treacle": Hooker, *Flora*, I:291.

camas roots: "Captain Lewis observes": Douglas, *Journal* (1914), 105.

"Its roots form a great part of the natives' food": Ibid.

A selection of plant specimens for William Jackson Hooker: Douglas to Hooker, December 1, 1828, Hooker Papers.

IV

"The Columbia presents a wide field for botanical research": Simpson, *Fur Trade*, 111.

"Do me the favour": Work, 99.

"I have all along experienced every attention in his power": Douglas, *Journal* (1914), 106.

"in compliance with your directions we have given every assistance to Mr. Douglas": McLoughlin (1941), 15.

"He expressed a desire": Ibid.

"I do not go alone": Douglas to Booth, October 2, 1825, Hooker Papers.

"We have been entertaining one another": Douglas, "Sketch," *CBM*, 133.

"an undertaking so arduous and interesting": Circular to HBC Officers, January 17, 1820, HBCA D.1/3, fo. 15.

"to make observations respecting the trade": Williams to John Clarke, January 12, 1820, HBCA D.1/3, fo. 6d.

"Tea is indispensable": Simpson, *Journal*, 261.

"like all the other Expeditions": Back, xxiii.

"not relish a botanist coming in contact with another's gleanings": Douglas, *Journal* (1914), 265.

Douglas meeting with John Richardson on the Saskatchewan: Ibid., 272.

"Mr. Garry was married on Tuesday last!": Douglas to Hooker, August 6, 1829, Hooker Papers.

"These gentlemen have much to contend with": Douglas to Hooker, May 6, 1834, Hooker Papers.

"I understand he is collecting materials for the press": Simpson to McLoughlin, June 28, 1836, HBCA D.4/22, fo. 38d.

"Company officers hadn't a soul above a beaver skin": Roberts, 191.

Grass widow
Sisyrinchium douglasii

Simpson considered Barnston "touchy": Simpson, Character Book, 230.

"he was hurried from a troubled seam": Barnston to Hargrave, February 2, 1836, Hargrave Papers.

V

"Finding two of the principal men who understood the Chenook tongue": Douglas, *Journal* (1914), 158.

"The Lion in the man awed all": MacDonald, 136–37.

McLoughlin family relationships: Watson, 2:673; von Kirk, 121, 155, 157.

Ogden-Rivet family: Watson, 2:734.

Finan McDonald: Ibid., 2:642.

"To this request I did not think proper to oppose myself": Connolly to Simpson, July 10, 1826. HBCA D.4/119.

"safely moored with her for life": Francis Ermatinger, 64.

nightcap "netted by an Indian girl": Douglas, *Journal* (1914), 249.

School of historians who address roles of mixed-blood women: See Barman, Brown, von Kirk, and Watson.

"But do you think David Douglas saw it?": Jackie Peterson, discussion with the author, January 2011.

"6 yards hair ribbon": Columbia Accounts, 1830, HBCA, B.223/d/32a, fo. 74.

"the son of David Douglas": Nancy Anderson, discussion with the author, January 2011.

Marie Josephte "Josette" Finlay: Courchane, discussion with the author, December 9, 2007.

David Finlay: Anderson to Simpson, April 17, 1850, HBCA D.5/28.

"a strong scent like mint": Douglas, *Journal* (1914), 179.

VI

"a very spirited boy": John Douglas, 1.

"no one could be more industrious and anxious to excel": Booth, Notice, 1.

"I like a devil better than a dolt": John Douglas, 5.

"contrived to make botany a first subject": Allan, 86.

"dreadfully knocked up and ill": Ibid., 81.

"voyaging to the ends of the earth": Allan, 56.

Hooker's "wandering demon": Ibid.

"David was invariably guided by the counsels": John Douglas, 8.

Background of the London Horticultural Society: Elliott, 1–11.

Lindley's collecting techniques: Lindley, 192.

"Much mischief being done to collections": Ibid., 195.

Douglas's collecting expedition to the mid-Atlantic region of the United States in 1823: Douglas, *Journal* (1914), 1–30.

"And here I cannot avoid adding my tribute": Richardson, "On Aplodontia," 337.

"begged me to name many of the plants": Hooker to Richardson, September 13, 1828, Hooker Papers.

"never seems so unhappy as when he has a pen in his hand": Ibid.

Douglas's "quarrelsome tendencies": Booth, Notice, 2.

Douglas's "outbreakings": Douglas to Hooker, June 7, 1829, Hooker Papers.

"He has much in his head": Hooker to Richardson, September 13, 1828, Hooker Papers.

"these will constitute a lasting memorial of Mr. Douglas's zeal": "List of Plants," *CBM*, 140.

VII

"a mere scrubby bush": Douglas, *Journal* (1914), 112.

Douglas on Garry oak: Douglas, "American Oaks," 48–49.

"Understanding which ecological processes maintained these oak stands": Dunwiddie, Environmental History, 7.

"to improve hunting conditions": Ibid., 8.

"Every fire is different": Dunwiddie, discussion with the author, March 20, 2011.

"everywhere fire destroyed all the grass": McLeod, October 2, 1826.

"in order that they might the better find wild honey": Douglas, *Journal* (1914), 214.

"parasitic on the roots of various grasses": Ibid., 135.

"Common on soils where wood has been destroyed by fire": Hooker, *Flora*, I:134.

Several oral accounts describe such stands as a primary benefit of such fires: Turner, "Time to Burn," 185–218.

Dunwiddie's sites and reconstituted suite of flowering plants: Dunwiddie, "Management and Restoration," 78–87.

Douglas's account of sugar pine: Douglas, "An Account of a New Species of *Pinus*," 497–500.

"dense gloomy forests": Douglas, "Some American Pines," 339.

"give to the mountain a peculiar—I was going to say an *awful*—appearance": Douglas to Hooker, November 20, 1831, Hooker Papers.

Douglas's account of Douglas-fir: Douglas, "Some American Pines," 338–42.

Douglas's descriptions of coniferous trees: Unless otherwise noted, all quotes are from Douglas, "Some American Pines."

"This part of the Columbia is by far the most beautiful that I have seen": Douglas, *Journal* (1914), 161.

"Delightful undulating country": Ibid., 170.

"a few yards from the river, in the shade of some pines": Douglas, *Journal* (1914), 173.

"What is missing": Rackham, correspondence with the author, March 17, 2011.

It is a semblance of such context that modern foresters are trying to recapture. Puettmann, *Managing*.

VIII

Unless otherwise noted, all Douglas quotes are from Douglas, *Journal* (1914), 152–57.

"The economy of animals": Swainson, *Taxidermy*, 1.

"for the express purpose of collecting plants": HBCA, A.6/21, fo. 11d.

"increased my collection of plants by seventy-five species": Douglas, "Sketch," *CBM*, 89.

Richardson's mammal identifications: Richardson, *Fauna*, 103, 172, 200.

Richardson's scientific paper on mountain beaver: Richardson, "On Aplodontia," 333–37.

Camas-rat quotes: Richardson, *Fauna*, 206.

Columbia sand-rat quotes: Ibid., 200–201.

Pocket gopher poisoning: Crouch, 662–64.

"Before parting with him I made inquiry": Douglas, *Journal* (1914), 172.

Evergreen violet
Viola sempervivens

"Here I purchased a pair of horns": Ibid., 210.

"The Horns are generally converted by the Snake Indians into bows": Douglas, "Observations on Two Undescribed Species," 382.

"The large hoof which this species has": Douglas, *Journal* (1914), 249.

IX

"the greatest scientific traveler": Barrett, 161.

"it would be interesting to compare": Douglas to Clinton, October 3, 1825, Clinton Papers.

Era of international cooperation: Multhauf, 11.

"While preparing for his departure": Sabine, Notes, 1.

"just so much knowledge of plane and spherical trigonometry": Ibid., 4.

"the relative intensity of magnetic attraction": Sabine, Observations, 146.

Sabine had developed a strict set of rules for taking readings: Ibid., 153–54.

sweeping "isographic" curves that covered the earth: Ibid., 147.

"capacity which enables him to take in knowledge": Sabine, Notes, 6.

"to undertake a variety of determinations": Ibid., 7.

"Mr. Douglas will be well provided with instruments": Sabine, Observations, 151.

Douglas's 1829–30 voyage around the Horn: Sabine, Notes, 8–9.

"His astronomical work advanced surely and rapidly": Barnston, 268.

"I have arranged my barometer every way to please me": Ibid., 274.

Douglas in California: Ibid., 275.

"Perhaps I may at a future time discuss . . . a treatise": Ibid.

"What a glorious prospect": Douglas to Hooker, August 6, 1829, *CBM*, 143.

"proceed northward, among the mountains": Douglas to Hooker, April 9, 1833, Hooker Papers.

Coordinates for New Caledonia: Sabine, Notes, 8–9.

"I am, thank God reconciled to my loss": Douglas to Hooker, May 6, 1834, Hooker Papers.

"Four feet below the extreme summit of the peak": Barnston, 331.

Letter to Edward Sabine: Douglas, "Volcanoes," 342–43.

Mauna Kea descriptions: Douglas, Ibid., 335–38.

X

"badger and rabbit holes" and "water very bad": Douglas, *Journal* (1914), 202.

"The species of *Calochortus* are bulbous rooted plants": Douglas, "An Account of the Species of *Calochortus*," 275.

"among the numerous vegetables used by them": Douglas, *Journal* (1914), 134.

Salal: Syd House, communication with the author, January 2011.

"strongly colonial": Hickman, 796.

"of no great edibility": National Museums Northern Ireland/ Habitas website (www. habitas.org.uk).

Felix Aripa: Discussion with the author, September 2010.

ILLUSTRATION AND CAPTION CREDITS

FRONT MATTER

Portrait of David Douglas
Daniel Macnee,1829
The Linnean Society, London
"Have you seen Douglas?":
  Tytler, 398.

Fair Journal
Royal Horticultural Society,
  London
Photograph courtesy of Lois
  Leonard.

PROLOGUE

Grand fir
Lambert, 1832
"The bark of the young trees":
Douglas, "Some American
  Pines," 343.

Douglas squirrel
Richardson, *Fauna*, vol. 1, plate
  14
"I procured some curious
  kinds":
Douglas, "Sketch," *CBM, 249.*

Blazing star
"Abundantly at the Great Falls
  of the Columbia":
Original Douglas specimen
  paper,
Natural History Museum,
  London.

Vasculum
Emily Nisbet
"I had in my pocket":
Douglas, *Journal* (1914), 15.

Snow douglasia
Edwards's Botanical Register,
  Vol. 22, t. 1886

"Upon close inspection":
*Quarterly Journal of Science,
  Literature and the Arts*, Vol.
  6, 1827.

CHAPTER 1

**Douglas Letter to DeWitt
Clinton**
"I soon found myself":
Clinton papers, Columbia
  University.

Albatross on Waves
Alfred Agate
Naval History Center,
  98-089-eq
"When rising from the water":
Douglas, *Journal* (1914), 53.

Ship in a Storm
Alfred Agate

Naval History Center,
98-089-gu
"The weather was so terribly
boisterous":
Douglas, *Journal* (1914), 55.

Heaving the Lead
John A. Atkinson
National Maritime Musuem,
PAD7766
"Dr. Scouler and I kept the
soundings":
Douglas, "Sketch," *CBM*, 241.

**North Head, Cape
Disappointment**
Photograph by John Marshall

Remnant of Shipwreck
Photograph by John Marshall
"Did you hear of the total
wreck":
Douglas to Hooker, September
14, 1829, Hooker Papers.

Cape Disappointment:
Henry James Warre
American Antiquarian Society
"The breeze improving":
Douglas, *Journal* (1914), 55.

**Chart of Columbia River Bar
1870**
NOAA Office of Coast Survey,
Historical Map & Chart
Collection 1130-00-1870
"The breadth of the Columbia":
Douglas, "Sketch," *CBM*,
242-43.

Lightship:
H. H. Hansen
National Maritime Museum
"Who was on board":
McDonald to Ermatinger,
February 20, 1833, Archives
of Canada.

CHAPTER 2

Encampment at Baker Bay:
Henry Eld
Beinecke Library, Yale
University
"Several canoes of Indians":
Douglas, *Journal* (1914), 102.

Salal
Edwards's Botanical Register,
Vol. 17, t. 1411
"On stepping on shore":
Douglas, *Journal* (1914), 102.

Chinook, Columbia River
Alfred Jacob Miller
Beinecke Library, Yale
University
"We met a number of Indians":
Scouler, "Journal," 163.

A Chinook Lodge c. 1846-47
Paul Kane (1810-1871)
watercolor and pencil on
paper, 5 1/8 x 7 1/4 inches,
Stark Museum of Art,
Orange Texas, 31.78.4.
"Cockqua, the principal chief of
the Chenooks":
Douglas, *Journal* (1914), 138.

Willapa Bay
Photograph by John Marshall

Chinook Tomb
Alfred Agate
Naval History Center,
NHC-98-089-gx
"An Intermittent Fever":
Douglas to Hooker, October 11,
1830, Hooker Papers.

Chinook cooking basket
Photograph by Mary Johnson
Family of Tony Johnson

"In the lodge were some
baskets":
Douglas, *Journal* (1914), 138.

Seashore lupine
Edwards's Botanical Register,
Vol. 14, t. 1198
"For eating, the roots are
roasted":
Douglas, *Journal* (1914), 138.

**Chinook Paddle, Made of
Oregon Ash**
Photograph by Mary Johnson
Family of Tony Johnson
"We had six Indians for
paddling":
Douglas, *Journal* (1914), 106.

Beargrass
Edwards's Botanical Register,
Vol. 19, t. 1613
"The natives at the Rapids":
Douglas, *Journal* (1914), 44.

Wrap Twined Basket
Photograph by Mary Johnson
Family of Tony Johnson

CHAPTER 3

View of Omak Lake
Alfred Downing 1882
Washington State Historical
Society
"Here, the whole country being
covered":
Douglas, *Journal* (1914), 161.

Digging Stick
Watercolor by Emily Nisbet

Douglas's Onion
Hooker, *Flora*, Vol. 2, Tab.
CSCVII
"A fine species of Allium":
Hooker, *Flora*, Vol. 2, p. 112.

Colville woman digging roots
University of Washington Library
"Roots are gathered in great quantities":
Hooker, *Flora*, Vol 1, p. 223.

Camas
Curtis's Botanical Magazine, Vol. 37, t. 1574
"Living on the roots":
Douglas, *Journal* (1914), 171.

Camas Meadow
Photograph by Charles Gurche

Douglas's brodiaea
Edwards's Botanical Register, Vol. 14, t. 1183

Purple Sage
Edwards's Botanical Register, Vol. 17, t. 1469
"At Priest's Rapid":
Hooker, *Flora*, Vol 2, p. 112.

Flat twined bag
Indian hemp and beargrass
Northwest Museum of Arts and Culture
"An Indian bag of curious workmanship":
Douglas, *Journal* (1914), 206.

CHAPTER 4

Field Telescope
Photograph by Denny DeMeyer

George Simpson
Hudson's Bay Company Archives, Archives of Manitoba, 363-S-25/5
"Had a note":
Douglas, *Journal* (1914), 294.

John McLoughlin
Courtesy of Old Oregon Photos

"Mr. McLoughlin kindly sent me":
Douglas, *Journal* (1914), 213.

Fort Vancouver
Henry James Warre
American Antiquarian Society
"The scenery from this place":
Douglas, *Journal* (1914), 56.

Temporary camp
Glenbow Archives
"I hastily bent my steps":
Douglas, *Journal* (1914), 191.

Traverse of the Saskatchewan
Henry James Warre
American Antiquarian Society
"Sundry articles gleaned":
Douglas, *Journal* (1914), 277.

Nicholas Garry
From a portrait in possession of his grandson
Hudson's Bay Company Archives, Archives of Manitoba, 1987/363-S-25/5
"Mr. Work showed me a pair of Mouton Blanche":
Douglas, *Journal* (1914), 246.

Wavy-leafed silk-tassel
Edwards's Botanical Register, Vol 20, t. 1686
"It forms a hardy evergreen shrub":
Hooker, *Flora*, Vol 2, p. 143.

Brown's Peony
Edwards's Botanical Register, Vol. 25, t. 30
"Flowers centre and the outside dark purple":
Douglas, *Journal* (1914), 192.

George Barnston
From a photograph in possession of his grandson

Heart-leaved buckwheat
Eriogonum compositum

Emily Nisbet
"I wish you had been with me":
Barnston, "Abridged Sketch," 271.

CHAPTER 5

Nisqually, Half-caste Indians
Henry James Warre 1845
American Antiquarian Society
"I was hailed into the camp":
Douglas, *Journal* (1914), 236.

Archibald MacDonald
Northwest Museum of Arts and Culture
"In company with Mr. Work":
Douglas, *Journal* (1914), 199.

Jane Klyne MacDonald
Northwest Museum of Arts and Culture

"After washing":
Douglas, *Journal* (1914), 209.

Douglas's Journal
Royal Horticultural Society, London
Photograph courtesy of Lois Leonard

Evergreen huckleberry
Edwards's Botanical Register, Vol. 16, t. 1354
"Faithful to his proposition":
Douglas, *Journal* (1914), 144.

Columbia Supply list
Hudson's Bay Company Archives, Archives of Mantiboa, B.223/d/32a

Josette Legace Work with children
British Columbia Archives, Victoria
"We were most cordially welcomed":
Douglas, *Journal* (1914), 246.

Indian Tobacco
Edwards's Botanical Register, Vol. 13, t. 1057
"I had seen only one plant before":
Douglas, *Journal* (1914), 141.

Sticky Currant
Edwards's Botanical Register, Vol. 15, t. 1263
"Mr. Finlay tells me":
Douglas, *Journal* (1914), 172.

Big-leafed Lupine
Edwards's Botanical Register, Vol. 13, t. 1377
"One of the most magnificent herbaceous plants":
Douglas, *Journal* (1914), 113.

CHAPTER 6

Royal Horticultural Society Medal
Oregon Historical Society
Photograph courtesy of Lois Leonard

Scone Palace, Perthshire
William Brown, *Select Views of Royal Palaces of Scotland*

William Jackson Hooker
T. Phillips, Engraved by H. Cook
National Library of Medicine, Scotland
"Let me repeat":
Douglas to Hooker, June 6th 1828, Hooker Papers

"Exhibitions Extraordinary in the Horticultural Room"
George Cruickshank

Plant Press
Lindley Herbarium, Kew Gardens

Douglas Journal
Royal Horticultural Society
Photo courtesy of Lois Leonard

Musk monkeyflower
Edwards's Botanical Register, Vol. 13, t. 1118
"A very beautiful plant":
Douglas, *Journal* (1914), 122.

Linnean Society Nomination
Linnean Society, London

Clarkia
Edwards's Botanical Register, Vol. 13, t. 1100
"Flowers rose color":
Douglas, *Journal* (1914), 131.

CHAPTER 7

Noble fir
Lambert, 1838
"This if introduced":
Douglas, "Some American Pines," 343.

Garry Oak
Thomas Nuttall, North American Silva
"Male flowers in pendulous, dense, hairy, yellow spikes":
Douglas, *Journal* (1914), 48.

Mt. Washington Willamette Country
Henry James Warre
American Antiquarian Society
"Country undulating; soil rich":
Douglas, *Journal* (1914), 213.

Garry Oak
Jeanne Debons
"Acorns, sessile in pairs":
Douglas, *Journal* (1914), 48.

Garry Oaks in Columbia Hills
Photograph by Charles Gurche
"Interspersed over the country":
Douglas, *Journal* (1914), 49.

Ground-cone
Hooker, Flora, Vol. 2, Tab. CLXVII.

Sugar Pine
Lambert, 1828
"At midday I reached my long-wished-for Pines":
Douglas, "Sketch," *CBM*, 89.

Douglas-fir
Forbes, *Pinetum woburnense*
"One of the most striking":
Douglas, "Some American Pines," 339.

Antelope bitterbrush
Purshia tridentata

Precision Dip Needle
Photograph by Dale Beeks
"Apparatus for Dip":
Douglas to Hooker, October 27,
   1829, Hooker Papers

Edward Sabine
Popular Science Monthly 2
   (1872): 238.
"Capt. Sabine kindly took
   Douglas":
Barnston, 267.

Magnetized compass needle
Photograph by Dale Beeks

Terrestrial Magnetism
Alexander Johnson, *Terrestrial
   Magnetism*, 1856
"The accompanying sketch of
   the northern hemisphere":
Sabine, Notes, 147.

Handheld Dip Needle
Photograph by Dale Beeks

Entering the Dalles
Henry James Warre
British Columbia Archives
"You may look upon":
Barnston, 270.

Bristlecone Fir
Lambert, 1838
"The singular thistle-like
   cones":
Douglas to Hooker, April 9,
   1833, Hooker Papers

Mountain barometer
"I have arranged my
   barometer":
Barnston, 271.

Walls of Crater
Joseph Drayton
Wilkes, *Narrative*, Vol. 4, p. 184
"At some places the hardened
   lava":
Douglas to Hooker, May 6,
   1834, Hooker Papers

CHAPTER 10

Sandstone and Basaltic Rocks
Henry James Warre
American Antiquarian Society
"From the oppressive heat":
Douglas, *Journal* (1914), 200.

Aspen Camp
James Madison Alden
National Archives,
   NWDNC-76-E221
"The whole country is
   destitute":
Douglas, *Journal* (1914), 202.

Sagebrush mariposa lily
Edwards's Botanical Register,
   Vol. 14, t. 1152

"I observed it in abundance":
Douglas, "An Account of the
   Species of *Calochortus*," 277.

**Specimen Paper of Splendid
mariposa lily**
The Board of Trustees of the
   Royal Botanical Gardens,
   Kew
"I have doubled the genus of
   Calochortus":
Douglas to Hooker, November
   20, 1831, Hooker Papers

Lambsquarters or Pigweed
Emily Nisbet
"Near Indian villages":
Douglas, *Journal* (1914), 194.

Cock of the Plains
Richardson, *Fauna*, Vol 2, p.
   358
"The pheasant-tailed grouse":
Douglas, "Observations of
   Some Species of the Genera
   Tetrao," p. 135-36.

Reed canarygrass
Emily Nisbet
"This does not seem to differ":
Hooker, *Flora*, Vol 2, p. 234.

Salmonberry
Edwards's Botanical Register,
   Vol. 17, t. 1198
"The young shoots":
Douglas, *Journal* (1914), 105.

Fireweed
Emily Nisbet
"From Lake Huron and
   Newfoundland":
Hooker, *Flora*, Vol 1, p. 221.

**Plateau Wetland with Reed
Canarygrass**
Photograph by Charles Gurche

BACK MATTER

158
Menzies' larkspur
Edwards's Botanical Register,
 Vol. 14, t. 1192

165
Grass Widow
Edwards's Botanical Register,
 Vol. 16, t. 1364

167
Evergreen violet
Edwards's Botanical Register,
 Vol. 15, t. 1254

171
Heart-leaved buckwheat
Edwards's Botanical Register,
 Vol. 21, t. 1774

173
Red-flowering currant
Edwards's Botanical Register,
 Vol. 16, t. 1349

174
Antelope bitterbrush
Edwards's Botanical Register,
 Vol. 17, t. 1446

177
Glacier lily
Edwards's Botanical Register.
 Vol. 21, t. 1786

179
Oregon iris
Edwards's Botanical Register,
 Vol. 15, t. 1218

181
Basketflower
Edwards's Botanical Register,
 Vol. 14, t. 1186

185
Douglas's brodiaea
Hooker, *Flora*, Vol. 2, Tab.
 CXCVIII

187
Richardson's Penstemon
Edwards's Botanical Register,
 Vol. 13, t. 1121

190
Showy Phlox
Edwards's Botanical Register,
 Vol. 16, t. 1351

BIBLIOGRAPHY

Allan, Mea. *The Hookers of Kew*. London: Michael Joseph Ltd., 1967.

Arno, Stephen F., and Ramona P. Hammerly. *Northwest Trees*. Rev. ed. Seattle: The Mountaineers, 2007.

Back, George. *Narrative of the Arctic Land Expedition to the Mouth of the Great Fish River, and Along the Shores of the Arctic Ocean in the Years 1833, 1834, and 1835*. London: John Murray, 1836. Facsimile. Reprint. Introduction by William C. Wonders. Edmonton: Hurtig, 1970.

Barman, Jean, and Bruce Watson. "Fort Colvile's Fur Trade Families & the Dynamics of Race in the Pacific Northwest," *Pacific Northwest Quarterly* 90 (Summer 1999): 140–53.

Barnston, George. "Abridged Sketch of the Life of Mr. David Douglas, Botanist, with a Few Details of His Travels and Discoveries." *Canadian Naturalist and Geologist* 5 (1860): 120–32, 200–208, 267–78, 329–49.

Barrett, Paul H., and Alain F. Corcos. "A Letter from Alexander Humboldt to Charles Darwin." *Journal of the History of Medicine and Allied Sciences* 27: 159–72.

Bell, Thomas. "Description of a New Species of Agama, Brought from the Columbia River by Mr. Douglass." *Transactions of the Linnean Society of London* 16 (1833): 105–7.

Booth, William Beattie. Memorandum on David Douglas. William Jackson Hooker Papers. Royal Botanic Gardens, Kew, England.

_____. Notice of the Early Life of the Late David Douglas. William Jackson Hooker Papers. Royal Botanic Gardens, Kew, England.

Boyd, Robert. *Indians, Fire, and the Land in the Pacific Northwest*. Corvallis: Oregon State University Press, 1999.

Brown, Jennifer S. H. *Strangers in the Blood: Fur Trade Company Families in Indian Country*. Vancouver:

University of British Columbia Press, 1980.

Camp, Pamela, and John G. Gamon, eds. *Field Guide to the Rare Plants of Washington*. Seattle: University of Washington Press, 2011.

Clinton, DeWitt. Papers. Columbia University Library, New York.

Courchane, David C. "The Descendants of James Finlay." In possession of the author.

Cox, E. H. M. "The Plant Collectors Employed by the Royal Horticultural Society, 1804–1846." *Journal of the Royal Horticultural Society* 80 (1955): 264–80.

Crouch, Glenn LeRoy. "Pocket Gophers and Reforestation in Western Forests." *Journal of Forestry* 80 (October 1982): 662–64.

Davies, John. *Douglas of the Forests*. Seattle: University of Washington Press, 1980.

Douglas, David. "An Account of a New Species of *Pinus*, Native of California." *Transactions of the Linnean Society* 15 (1827): 497–500.

———. "An Account of Some New, and Little Known Species of the Genus *Ribes*." *Transactions of the Horticultural Society* 8 (1830): 509–18.

———. "An Account of the Species of *Calochortus*; a Genus of American Plants." *Transactions of the Horticultural Society* 7 (1830): 275–80.

———. "American Oaks." In *Journal Kept by David Douglas, 1823–1827*. London: William Wesley & Son, 1914.

———. "Description of a New Species of the Genus *Pinus (P. Sabiniana)*. *Transactions of the Linnean Society* 16 (1833): 747–50.

———. Field Sketches, Fort Okanagan to Quesnel River, 1833. British Columbia Archives, Victoria.

———. *Journal Kept by David Douglas During His Travels in North America*. London: William Wesley and Son, 1914.

———. "Observations of Some Species of the Genera *Tetrao* and *Ortyx*." *Transactions of the Linnean Society* 16 (1833): 133–49.

———. "Observations on the *Vultur californianus* of Shaw." *Zoological Journal* 4 (1829): 328–30.

———. "Observations on Two Undescribed Species of North American Mammalia." *Zoological Journal* 4 (1829): 330–32.

Glacier lily
Erythronium grandiflorum

———. Papers. Lindley Library. Royal Horticultural Society, London.

———. "A Sketch of a Journey to the North-Western Parts of the continent of North America, during the Years 1824, 5, 6, and 7." *Companion to the Botanical Magazine* 2 (1836): 82–177.

———. "Sketch of a Journey to North-West America, 1824–27." In *Journal Kept by David Douglas, 1823–1827*. London: William Wesley & Son, 1914.

———. "Some American Pines." In *Journal Kept by David Douglas, 1823–1827*.

London: William Wesley & Son, 1914.

———. "Volcanoes in the Sandwich Islands." *Geographical Society Journal* 4 (1834): 333–43.

Douglas, John. Biographical Notes of David Douglas. William Jackson Hooker Papers. Royal Botanic Gardens, Kew, England.

Dunwiddie, Peter. Environmental History of a Garry Oak/Douglas-fir Woodland on Waldron Island, Washington. In press, 2012.

———. "Management and Restoration of Grasslands on Yellow Island, San Juan Islands, Washington, U.S.A." *Proceedings of the Third Annual Meeting of the B.C. Chapter of the Society for Ecological Restoration*, April 2002.

——— and Jonathan D. Bakker. "The Future of Restoration and Management of Prairie-Oak Ecosystems in the Pacific Northwest." *Northwest Science* 85 (May 2011).

Elliott, Brent. *The Royal Horticultural Society: A History, 1804–2004.* Chichester, England: Phillimore & Co., 2004.

Ermatinger, Edward. "Edward Ermatinger's York Factory Express Journal . . .

1827–28." Edited by C. O. Ermatinger and James White. *Proceedings and Transactions of the Royal Society of Canada*, 3rd ser., 6 (1912): 67–132.

Ermatinger, Francis. *Fur Trade Letters of Francis Ermatinger.* Edited by Lois Halliday McDonald. Glendale: The Arthur H. Clark Company.

Fleming, J. H. "The California Condor in Washington." *The Condor* 26 (May 1924): 111–12.

Fletcher, Harold R. *The Story of the Royal Horticultural Society, 1804–1968.* London: Oxford University Press, 1969.

Fort Vancouver Account Books. Hudson's Bay Company Archives, Winnipeg, Manitoba.

Fort Vancouver Correspondence. Hudson's Bay Company Archives, Winnipeg, Manitoba.

Franklin, John. *Narrative of a Journey to the Shores of the Polar Sea in the Years 1819-20-21-22.* London: John Murray, 1824.

Gairdner, Meredith. Correspondence. William Jackson Hooker Papers. Royal Botanic Gardens, Kew, England.

Gunther, Erna. *Ethnobotany of Western Washington:*

The Knowledge and Use of Indigenous Plants by Native Americans. Seattle: University of Washington Press, 1973.

Hall, F. S. "Studies in the History of Ornithology in the State of Washington (1792–1932) with Special Reference to the Discovery of New Species: Part III." *The Murrelet* 15 (January 1934): 2–19.

Hargrave, James. James Hargrave and family fonds. Library and Archives Canada, Ottawa, Ontario.

Harvey, Athelstan G. *Douglas of the Fir: A Biography of David Douglas, Botanist.* Cambridge, MA: Harvard University Press, 1947.

Hickman, James C., ed. *The Jepson Manual: Higher Plants of California.* Berkeley: University of California Press, 1993.

Hitchcock, C. Leo, and Arthur Cronquist. *Flora of the Pacific Northwest: An Illustrated Manual.* Seattle: University of Washington Press, 1973.

Hooker, Joseph Dalton. *A Sketch of the Life and Labours of Sir William Jackson Hooker, Late Director of the Royal Gardens of Kew.* Cambridge: Cambridge University Press, 1903.

Hooker, William Jackson. "A Brief Memoir of the Life of Mr. David Douglas, with Extracts from His Letters." *Companion to the Botanical Magazine* 2 (1836): 79–182.

_____. *Flora Boreali-Americana; being the Botany of the Northern Parts of British America.* 3 vols. London: Henry G. Bohn, 1829–40.

_____. "List of Plants Introduced by Mr. Douglas." *Companion to the Botanical Magazine* 2 (1836): 140–42.

_____. Papers. Royal Botanic Gardens, Kew, England.

Hudson's Bay Company Archives, Winnipeg, Manitoba.

Hunn, Eugene. *Nch'I=Wana: The Big River: Mid-Columbia Indians and Their Land.* Seattle: University of Washington Press, 1990.

Lambert, Aylmer Bourke. *A Description of the Genus Pinus.* 3 vols. London: Weddell, Prospect Row, Walworth, 1803, 1828, 1832.

Lindley, John. "Instructions for Packing Live Plants in Foreign Countries." *Transactions of the Horticultural Society of London* 6 (1826): 293–95.

Log of the *Willliam and Ann,* 1824–25. Hudson's Bay Company Archives, Winnipeg, Manitoba.

Lyell, Charles. *Principles of Geology.* Vol. 1. London: John Murray, 1830.

McDonald, Archibald. Papers. British Columbia Archives, Victoria.

_____. *This Blessed Wilderness: Archibald McDonald's Letters from the Columbia, 1822–44.* Edited by Jean Murray Cole. Vancouver: UBC Press, 2001.

McDonald, Ranald. *Ranald McDonald, the Narrative of His Early Life.* Edited by William S. Lewis and Naojiro Murakami. Spokane: Eastern Washington Historical Society, 1923.

McLeod, Alexander. Journal. Hudson's Bay Company Archives, Winnipeg, Manitoba.

McLoughlin, John. *Letters of Dr. John McLoughlin Written at Fort Vancouver 1829–32.* Edited by Burt Brown Barker. Portland, OR: Binford & Mort, 1948.

_____. *Letters of Dr. John McLoughlin Written at Fort Vancouver to the Governor and Committee.* Edited by E. E. Rich. London: The Hudson's Bay Record Society, 1941.

Meyers, J. A. "Finan McDonald—Explorer, Fur Trader, and Legislator."

Oregon iris
Iris tenax

Washington Historical Quarterly 13 (1922): 196–208.

Mitchell, Ann Lindsey, and Syd House. *David Douglas: Explorer and Botanist.* London: Arum Press, 1999.

Moulton, Gary, ed. *The Journals of the Lewis & Clark Expedition.* 12 vols. Lincoln: University of Nebraska Press, 1983–2001.

Multhauf, Robert P. and Gregory Good. *A Brief History of Geomagnetism and A Catalog of the Collections of the National Museum of American*

History. Washington, DC: Smithsonian Press, 1987.

Nisbet, Jack. *The Collector: David Douglas and the Natural History of the Northwest.* Seattle: Sasquatch Books, 2009.

———. *The Mapmaker's Eye: David Thompson on the Columbia Plateau.* Pullman: Washington State University Press, 2005.

———. *Sources of the River: Tracking David Thompson across Western North America.* Seattle: Sasquatch Books, 1994, 2007.

———. *Visible Bones: Journeys across Time in the Columbia River Country.* Seattle: Sasquatch Books, 2003.

Nuttall, Thomas. *The Genera of North American Plants, and a Catalogue of the Species to the Year 1817.* Philadelphia: Printed for the Author by D. Heartt, 1818.

Oliver, F. W., ed. *Makers of British Botany.* Cambridge: University Press, 1913.

Peattie, Donald Culross. *A Natural History of Western Trees.* New York: Bonanza Books, 1950.

Puettmann, Klaus J., K. David Coates, and Christian Messier. *A Critique of Silviculture: Managing for Complexity.* Washington, DC: Island Press, 2008.

Rackham, Oliver. *Woodlands.* London: HarperCollins Publishers Ltd., 2006.

Rich, E. E. *History of the Hudson's Bay Company, 1670–1870.* 2 vols. London: Hudson's Bay Record Society, 1960.

Richardson, John. Correspondence. William Jackson Hooker Papers. Royal Botanic Gardens, Kew, England.

———. *Fauna Boreali-Americana, or the Zoology of the Northern Parts of British America.* 2 vols. London: John Murray, 1829.

———. "On Aplodontia, a New Genus of the Order Rodentia." *Zoological Journal* 4 (1829): 333–37.

Roberts, George B. "Letters to Mrs. F. F. Victor." *Oregon Historical Quarterly* 63 (1962): 175–205.

Ruby, Robert H., and John A. Brown. *The Spokane Indians: Children of the Sun.* Norman: University of Oklahoma Press, 1970.

Sabine, Edward. Notes on the Geographical Observations Made by David Douglas on the Pacific Coast of North America. 1837. British Columbia Archives, Victoria.

———. Observations Taken on the Western Coast of North America, by the Late Mr. David Douglas. British Columbia Archives, Victoria.

———. "Report on the Variations of Magnetic Intensity Observed at Different Points of the Earth's Surface." *Report of the Seventh Meeting of the British Association for the Advancement of Science* 6 (1837): 1–85; 500.

Scouler, John. "Journal of a Voyage to N.W. America." *Oregon Historical Quarterly* 6 (1905): 54–75, 159–205, 276–87.

———. "Observations on the Indigenous Tribes of the N.W. Coast of America." *Journal of the Royal Geographical Society of London* 11 (1841): 215–51.

———. "Remarks on the Form of the Skull of the North American Indian." *Zoological Journal* 4 (1829): 304–9.

Simpson, George. Character Book. Hudson's Bay Company Archives, Winnipeg, Manitoba.

———. Correspondence. Hudson's Bay Company Archives, Winnipeg, Manitoba.

———. *Fur Trade and Empire: George Simpson's Journal, 1824–25.* Edited by Frederick Merk. Cambridge, MA: Belknap Press of

Harvard University Press, 1968.

_____. *Journal of Occurrences in the Athabasca Department.* Edited by E. E. Rich. Toronto: Champlain Society, 1938.

Suttles, Wayne, and William Sturtevant, eds. *Handbook of North American Indians.* Vol. 7. *Northwest Coast.* Washington, DC: Smithsonian Institution, 1998.

Swainson, William. *Taxidermy; with the Biography of Zoologists.* London: Longman, 1840.

_____. *The Naturalist's Guide for Collecting and Preserving Subjects of Natural History and Botany.* London: W. Wood, 1822.

Tolmie, William Fraser. *The Journals of William Fraser Tolmie, Physician and Fur Trader.* Vancouver, BC: Mitchell Press Limited, 1963.

_____. Papers. British Columbia Archives, Victoria.

Transactions of the Horticultural Society of London, 1824–34.

Turner, Nancy. *Food Plants of Interior First Peoples.* Victoria: Royal British Columbia Museum, 1997.

_____. "Time to Burn." In *Indians, Fire, and the Land in the Pacific Northwest,* edited by Robert Boyd, 185–219. Corvallis: Oregon State University Press, 1999.

_____, Randy Bouchard, and Dorothy Kennedy. *Ethnobotany of the Okanagan-Colville Indians of British Columbia and Washington.* Occasional Papers of the British Columbia Provincial Museum, no. 21, 1980.

Tytler, Patrick Fraser, and James Wilson. *Historical View of the Progress of Discovery on the More Northern Coasts of America.* Edinburgh: Oliver & Boyd, 1832.

Von Kirk, Sylvia. *Many Tender Ties: Women in Fur-Trade Society in Western Canada, 1670–1870.* Winnipeg, Manitoba: Watson & Dwyer Publishing, 1981.

Walker, Deward E. Jr., ed. *Handbook of North American Indians.* Vol. 12, *Plateau.* Washington, DC: Smithsonian Institution, 1998.

Watson, Bruce. *Lives Lived West of the Divide: A Biographical Dictionary of Fur Traders Working West of the Rockies 1793–1858.* 3 vols. Kelowna: University of British Columbia, Okanagan, Centre for Social, Spatial, and Economic Justice, n.d.

Basketflower
Gaillardia aristata

Wilson, James. *Illustrations of Zoology.* Edinburgh: William Blackwood, 1831.

Work, John. "The Journal of John Work." Edited by T. C. Elliott. *Washington Historical Quarterly* 6 (1915): 26–49.

INDEX

Douglas's brodiaea
Tritelia grandiflora

Richardson's penstemon
Penstemon richardsonii

Showy phlox
Phlox speciosa

ABOUT THE AUTHOR

Teacher and naturalist **Jack Nisbet** is the author of several books that explore the human and natural history of the Intermountain West, including *Purple Flat Top*, *Singing Grass Burning Sage*, and *Visible Bones*. *Sources of the River* and *The Mapmaker's Eye* trace the adventures of pioneering fur agent and cartographer David Thompson, and *The Collector* follows David Douglas's forays through the New World. To find out more, visit www.JackNisbet.com.